MARS ONE

HUMANITY'S NEXT GREAT ADVENTURE

HUMANITY'S NEXT GREAT ADVENTURE

INSIDE THE FIRST HUMAN SETTLEMENT ON MARS

EDITED BY NORBERT KRAFT, MD
WITH JAMES R. KASS, PHD,
AND RAYE KASS, PHD

BenBella Books, Inc.
Dallas, Texas

BenBella Books, Inc.
10300 N. Central Expressway, Suite #530 | Dallas, TX 75231
www.benbellabooks.com
Send feedback to feedback@benbellabooks.com

Printed in the United States of America
10 9 8 7 6 5 4 3 2 1

Library of Congress Cataloging-in-Publication Data
Names: Kraft, Norbert (Physician), editor. | Kass, James R., editor. | Kass, Raye., editor.
Title: Mars One, humanity's next great adventure : inside the first human settlement on Mars / edited by Norbert Kraft ; with James R. Kass and Raye Kass.
Description: Dallas, TX : BenBella Books, Inc., [2016] | Includes bibliographical references and index.
Identifiers: LCCN 2015039198| ISBN 9781940363837 (trade paper : alk. paper) | ISBN 9781940363936 (electronic : alk. paper)
Subjects: LCSH: Mars (Planet)—Exploration.
Classification: LCC QB641 .M3636 2016 | DDC 919.9/2304—dc23 LC record available at http://lccn.loc.gov/2015039198

Copyediting by Francesca Drago
Proofreading by Lisa Story and Michael Fedison
Text design by Silver Feather Design
Text composition by PerfecType, Nashville, TN
Cover design by Sarah Dombrowsky
Cover illustration by Bryan Versteeg, Mission Concept Artist, Mars One
Outpost illustrations courtesy Bryan Versteeg/Mars One
Printed by Lake Book Manufacturing

Distributed by Perseus Distribution
www.perseusdistribution.com

To place orders through Perseus Distribution:
Tel: (800) 343-4499
Fax: (800) 351-5073
E-mail: orderentry@perseusbooks.com

Permissions

ACKNOWLEDGMENTS

The editors wish to express their gratitude to all the Mars One candidates whose quotes have been used in these pages, including the following, who contributed thoughts to "Age and Aging on Mars": **Dan Carey**, **Ethan Dederick**, **Yuri Rafael Lopez Farias**, **Reginald Foulds**, **Laurel Helene Kaye**, **Mead McCormick**, **Dianne McGrath**, **Elaheh Nouri**, **Sue Ann Pien**, **Jaymee Orillosa Del Rosario**, **Etsuko Shimabukuro**, **Dr. Bhupendra Singh**, **Anastasiya Stepanova**, and **Kay Radzik Warren**. We regret that time did not allow us to contact more. Interviews were conducted by Mars One Exchange writer **Vincent Hyman**, who also wrote up the interview results for this chapter.

The editors wish to thank **Leah Wilson** (editor–in-chief) and **Jessika Rieck** (production associate) at BenBella Books for their hard work and patience throughout the journey this book took from draft to final copy.

The editors would also like to thank **Stephen Cass**, senior editor with *Technology Review* and founding editor of *Discover* magazine's sci-fi blog Science Not Fiction, for his insightful advice and feedback on all things technological (and others too numerous to mention).

The editors would like to extend their profound thanks to **Suzanne Flinkenflögel**, Director of Communications for Mars One, whose expertise and incredible skill has made this book possible.

Thanks are due as well to **Bryan Versteeg**, the illustrator responsible for all the illustrations and animations on the Mars One website, for the outpost images used in this book. Thanks also to **Olga Panifilova** for creating the press release and advertisement images in "A World Waiting to Be Born."

Finally, the editors would like to especially acknowledge **Bas Lansdorp**, cofounder and chief executive officer, and **Arno Wielders**, cofounder and chief technical officer, of Mars One. Without their innovation and willingness to push boundaries due to their passion for science and exploration, this book would have remained only in the minds and hearts of those whose words follow. Because of their vision of what could be, this book exists.

CONTENTS

FOREWORD

A young man appeared in my office. Yes, he had made an appointment with my secretary to meet me. He had heard about my interest in futuristic applications of science and science fiction. He, and a few companions, had some rather bold ideas that he wanted to unfold for me, and he wanted to learn how I would react.

If you are in a field of science like mine, people with "new ideas" show up all the time. Those ideas are usually "new" and "original" indeed, but completely out of touch with the real world. Usually, they are based on hopelessly ill-informed perceptions of what real science and technology are about, and there is not much I can do for such people other than advise them to learn much more about what professionals have to say regarding the topics they are so thrilled about before bothering me again.

And here was a guy talking about human colonies on the planet Mars. The colonists would travel for approximately seven months from Earth to Mars, four people at a time, and they would stay there, keeping themselves alive as long as they were able, without the slightest glimmering of hope for a safe return home to Earth but ensured of eternal fame in the history of mankind.

The date: April 27, 2012.

The man's name: Bas Lansdorp.

"All basic science you need already exists," he told me. "All you need to do is some more research to combine it all. Use spacecraft designs that have already been made, scale them up a bit, and test them thoroughly." It would take just about ten years to prepare for the first colonists to go, according to his calculations, and the cost would be around 6 billion American dollars.

Now this, I thought, was entirely unrealistic. It's true that basic science did not put any fundamental obstacles in his way. In principle, all this was possible. But those numbers? Ten years? Six billion dollars? "You better put an extra zero behind them," I said. Building a big particle accelerator already costs more, as do large railway projects, let alone the development of a new military aircraft.

But he was sure he had done his calculations correctly. Yes, his estimates were optimistic; if things did not go as planned, it would take longer and cost more, he admitted. "But we will be a private enterprise. We will not depend on government declarations and political hassles. We'll make our decisions swiftly."

It still sounded very optimistic indeed. But when hearing a thing like this, I do not only listen to how it sounds. I also search for grave mistakes and gross misperceptions. Here, these were lacking. It seemed that he had considered all major issues involved. How do people survive for seven months in a spaceship to Mars? How does a spaceship with people in it land on Mars? And, after they have landed, how do you provide them with water, energy, food, and a lasting atmosphere? How should they communicate with mission control on Earth? How do you protect them against major hazards, such as cosmic radiation and energetic solar bursts, but also more mundane things such as poison and dust? How will the colonists handle numerous threats, such as malfunctioning devices, medical problems, life-threatening shortages of almost any one of their basic needs, and so on?

Not only had he considered those concerns, but already plans had been made for assembling the funds needed for the project. "It will be a unique accomplishment for humanity. Cash will flow," he claimed.

His answers were optimistic but not crazy. Indeed, in principle, one can take care of all major aspects of journeys to Mars. Then something happened that I had not anticipated: I became enthusiastic about his ideas. If it is at all possible, humans should be enabled to travel to Mars, though I still think that it will take longer, perhaps much longer, than ten years before a manned spaceship can lift off toward Mars, and that the cost will be more than the estimated 6 billion.

If man can go to Mars and establish settlements there, then the rest of the solar system comes into view. What about the moon, some of the larger asteroids, the large moons of Jupiter and Saturn? What about the cold, remote regions of the outer solar system?

"I will be your Ambassador," I said, "but I won't cover the financial part. I never did understand such calculations." Because if it can be done, it doesn't matter how long it takes. I see a multitude of good and interesting developments that could be triggered by this activity. Even if Mars One never succeeds in putting people on Mars itself, the organization will constantly arouse people's imagination about Mars, trigger new investigations and investments in all those technical issues associated with manned Mars missions, and otherwise pave the road in many ways. Humanity will go to Mars, even if we cannot yet say exactly when and how.

Naturally, there are still many questions to be answered. One essential ingredient of the plan is that unmanned machines will precede the manned missions. Robots will be employed to do as much as possible before the colonists arrive. Robots? What can they do? Does Mars One realize how difficult it is to persuade robots to do just about anything? These will have to be intelligent machines, and they don't exist yet. This is going to be a tough problem. Then, when they arrive, the colonists will be surrounded by an extremely hostile environment, dependent on highly advanced technology as much as their own ingenuity. Much of their habitat will have to be covered with thick layers of earth for protection.

Often, as well, the question is raised: What good will sending people to Mars be for science? Many scientists regard manned spaceflight as a ludicrous waste of money and effort. Scientific exploration can be done much better and much cheaper by robots, they emphasize. Indeed, this is true for most scientific questions being investigated now. But what is it that makes planets and moons so interesting in the first place? Could it be related to the remote possibility that humans might someday set up camp there?

To my mind, manned exploration of the distant parts of the solar system would be one marvellous scientific experiment all in itself. Can a lasting biological ecosystem be set up on places other than Earth, places where such ecosystems do not exist at present? How will such systems evolve there, together with the species *Homo sapiens*? Can they, in unison, defend themselves against numerous external threats? Will humans be strong enough to persevere, even if mishaps do take place?

How far can we go? Or will the task of colonizing these places be taken over by intelligent robots altogether?

I am Mars One's ambassador because this attempt is the first of its kind. It is going considerably further than any of its more amateuristic predecessors. Maybe it won't succeed in its primary mission, but whatever errors it makes now will be extremely instructive for the next initiatives. In that case, we can still maintain that Mars One is paving the road to the future.

This book is one of the instruments for doing so. It discusses the skills human colonists will need to comply with all conceivable eventualities that they may encounter. Colonists will need not only suitable air, water, food, and shelter, but also the strength to endure complications in their social relationships, challenging workloads, loneliness perhaps, and other sufferings. They will need resistence and perseverance.

One thing they will enjoy for sure: the first colonists will have earned fame and the admiration of the millions back on Earth watching them, much like Olympic athletes. Their successors on Mars, also, will become famous for their attempts to expand and strengthen their colonies. They all will be the ones who did it.

—Gerard 't Hooft

INTRODUCTION

We have been orbiting humans around Earth for more than half a century. We have landed on the moon and sent spacecraft to explore various near and distant planetary objects in our solar system. The next logical place for humanity's great adventure is Mars. It's the only planet we know of with an abundance of a key life-critical substance—water—and although located much farther away from Earth than our moon, it is close enough for humans to make the trip, even with our current technology.

Mars One is a nonprofit organization, based in the Netherlands and international in scope, whose goal is to establish a permanent human settlement on Mars. Why do this? Because it is the next giant leap forward for humankind; a stepping-stone for the human race on its unyielding quest to explore the universe. Human settlement on Mars will aid our understanding of the origins of the solar system, the origins of life, and our place in the cosmos.

Sending a manned mission to Mars is a fantastic adventure. Imagine the incredible feeling of being the first human in history to step out of the capsule and leave your footprint on the surface of Mars. This feeling of amazement will be experienced by not only the astronaut but also by his or her audience: all those watching from back home. After all, many of those who observed Neil Armstrong land on the moon so many years ago still remember the details—where they were, who they were with, and how they felt when it happened. This will be our moment, in 2027. Imagine, too, something more—what it would be like to live on another planet, millions of miles from Earth, and look up into the night sky, knowing that one of the "stars" is actually the planet on which you were born.

And what about humankind's good old-fashioned curiosity? What can studying Mars teach us about Earth's history? Is there viable life already living on the Red Planet? These are a few of the many burning questions for scientists worldwide to ponder and seek to answer.

Progress is another reason to establish a permanent human settlement on Mars. A mission to Mars will jump-start massive developments in areas such as recycling, solar energy, food production, and medical technology, to name just a few.

Next to planet Earth, Mars is the most habitable planet in our solar system. Its soil contains water to extract; it isn't too cold or too hot; there is enough sunlight to power solar panels. Gravity on Mars is 38 percent that of our Earth and, although that may sound low, many believe it is a sufficient amount for the human body to be able to adapt to. Mars has an atmosphere (albeit a thin one) that offers protection from cosmic and the sun's radiation. Its day-night rhythm is very similar to Earth's: A Mars day is twenty-four hours, thirty-nine minutes, and thirty-five seconds long.

In contrast, consider the only other two celestial bodies in orbit near Earth: our moon and Venus. There are far fewer vital resources available on the moon, and a moon day lasts one month. It also does not have an atmosphere to form a barrier against radiation. Venus is a veritable purgatory. The average temperature is more than 400 degrees, the barometric pressure is the same as if you were located 900 meters (about one-half nautical mile) underwater on Earth, and there are occasional bouts of acid rain. Venus has nights that last for 120 days. Humans cannot live on Mars without the help of technology, but compared to Venus it's paradise!

As of this writing, Mars One's first unmanned mission to Mars, to prepare a habitable settlement, is scheduled to depart in 2020. The first four-person crew will depart for its one-way journey to Mars starting in 2026. But the arrival of the first four to inhabit Mars is just the beginning of this great adventure. Those astronauts will be followed by subsequent crews, which will depart every twenty-six months thereafter, leading the Martian base to grow eventually into a small village. At first, expansion will be limited due to provisions, oxygen, and water. Other landings will provide everything the settlers might need to expand the colony: new living quarters, solar panels, and plastic components. The settlement will continue to develop as those inhabiting it become architects of their own environment.

After Mars One's official announcement, a large number of candidates registered their interest in being a member of the first team to settle on Mars. In 2013, nearly a quarter-million applicants filled out Mars One's online registration. As of February 2015, that number was whittled down to one hundred, and, in 2016, the selection process will reduce that number to twenty-four candidates—those who will receive a formal, full-time employment offer followed by a ten-year training schedule for Mission Mars. Both the selection and training process will be filmed and shared with the world. (To read more about the selection process—the journey from the original applicant pool to the remaining one hundred and the work that remains—please see "The Mars One Selection Process" on page 257.)

In this anthology, we, the editors and core members of the Mars One Selection Committee, aim to answer a number of questions about the requisite skills the members of this first crew to Mars must have, along with some reflections about the effects that continuous filming may have on them. We also share a selection of answers the applicants gave about their lives on Earth and how they may potentially live their lives with others on Mars.

To answer these questions, we've divided this book into four main parts to which well-known experts—from former NASA employees to multi-decade veterans in the spaceflight field—have contributed their learned opinions, providing valuable information and insights about the unknown variables the settlers could face across the many possible scenes of this challenging adventure.

Part I discusses the "hard" scientific skills, falling across a broad range of domains, the crew members will be expected to possess:

technical expertise for maintaining the spacecraft and the habitat, along with the aptitude and inclination to innovate solutions to unforeseen problems, with limited tools and supplies; and the knowledge, techniques, and confidence needed to be able to provide their own medical care, when necessary, and to keep healthy and fit.

Part II discusses the so-called soft skills that are critical for facilitating team cohesion despite the complex interplay of interpersonal dynamics among multicultural, gender-mixed crew members who may differ widely in age.

Part III looks at the unique circumstances in which the crew will be learning and practicing these skill sets—in front of the cameras, with their actions projected to viewers worldwide—and at the impact continuous filming could have on the candidates and their audience.

Part IV imagines a typical workday on Mars: What would it be like? How might one's leisure time be spent afterward? We'll also touch on the political and legal complexities that surround colonizing another planet.

One thing is certain: Human beings will visit the planet Mars—orbit around it, land on it, and maybe even settle there—and we have to be prepared for this. Within this anthology, we, the editors and core members of the Mars One Selection Committee, will take you through the key stages of the Mission Mars adventure, and we hope you will not only enjoy reading about them but also become better informed about the multifaceted dimensions of fulfilling this challenge!

—*The Editors*

MARS ONE
HUMANITY'S NEXT
GREAT ADVENTURE

Technical and Medical Skills, Health and Fitness

Four crew members are in a spaceship on their way to Mars. They have been traveling for some months, and already Earth has shrunk to the size of a golf ball. The trip has been almost boring, going well and without incident. The crew members have passed their time maintaining the ship, taking care of themselves, and reviewing the various emergency procedures.

Before leaving Mother Earth, they had been briefed on and trained through a large number of contingencies devised by both the ground experts and themselves. Of course they knew that not all contingencies could be foreseen, but they had gone through much technical training on the complex ship, which, theoretically, they now knew inside out—well, more or less, because no one could know everything, and not everything inside the ship was accessible to them. But all had gone well—until now.

Suddenly, the lights in the spaceship start flickering and several warning lights begin blinking. A small meteorite has just hit the spaceship, probably damaging an array of solar panels.

The crew immediately starts the emergency procedures for which it has been well trained. Switching operations to low power, crew members quickly check for a leak—are they losing pressure, and if so, where? What's the most effective, most expedient way to repair it?

What are the skills the Martian travelers will need in order to cope with situations, foreseen and unforeseen, like this one? The answer is—many! As the authors in this section point out, they will not necessarily need PhDs, but they will *need certain practical skills in order to not just survive but thrive in their new environs.*

This section of the anthology addresses several "hard" science skills that will be required by the Mars One crew members (soft skills are addressed in another section). We have selected three skill areas that we consider of paramount importance: technical, to ensure crew members are able to maintain, repair, innovate, and build necessary equipment; medical, to provide routine checkups, preventive and emergency care, and even perform small operations; and health and fitness, to sustain excellent performance, maintain good health, and keep fit.

Our first contributor, Mars One advisor Dr. Mason Peck, is a professor in the Sibley School of Mechanical and Aerospace Engineering at Cornell University and has served as Chief Technologist for NASA, as well as in other key posts in the aerospace industry. Peck commands a broad background in aerospace technology, which comes from nearly twenty years of research and development work in academia and industry.

In his essay "Improvisation and Exploration," Dr. Peck underscores the human element in overcoming technical challenges for survival on Mars. Drawing on his lengthy experience in the domain of spaceflight, he describes unusual examples of how unforeseen problems were solved and postulates on the types of potential tasks that await the first settlers on Mars, as well as the related skills and tools they will need to successfully carry out those tasks.

Next up is Mars One advisor Dr. Thais Russomano, a medical doctor with more than twenty years of experience in aerospace medicine, space physiology and medicine, biomedical engineering, and telemedicine and eHealth research and development.

Dr. Russomano begins her essay "Medical Skills for an Interplanetary Trip" with a discussion of the hostile environments that exist both in space and on the planet Mars, their potential negative effects on human bodies, and the preventative and corrective steps that can ameliorate these effects. She outlines the skills the crew will need to monitor their health both on the journey to Mars and during daily life on the planet, the facilities required, and the telemedicine-diagnoses scenarios with experts from Earth who could support these activities—leaving the reader with a feeling of great thankfulness for the environmental conditions we enjoy on our own planet Earth!

Finally, we hear from Jamie R. Guined who, just as Drs. Peck and Russomano, acts as advisor to Mars One and, moreover, currently serves as program manager and a commercial scientist-astronaut for Project PoSSUM (Polar Suborbital Science in the Upper Mesosphere), the world's first manned *commercial suborbital research program. Guined also previously served as an exercise scientist (kinesiologist) with the University of Houston, Texas, supporting the Neurosciences Laboratory at the Johnson Space Center.*

In her essay "Human Health and Performance for Mars Missions," Guined points out that there are still some outstanding gaps in knowledge regarding human health and performance risks for a mission to Mars. She expounds first on the physiological considerations of the giant leap from low Earth orbit missions to a Mars surface mission, then recounts in a nutshell the history of exercise countermeasures in human spaceflight from early days through the NASA Skylab, Soviet Salyut, Mir, and Shuttle programs. Finally, regarding a mission to the Mars surface, she considers two distinct stages with their own special needs: (1) the maintenance of human health and performance during Mars transit, and (2) the first thirty days on Mars itself, and the concomitant readaptation to gravity and reconditioning. Guined closes with a summary of the key considerations for establishing a preventative exercise countermeasures program for permanent Mars habitation.

—The Editors

What technical skills will the crew need to survive and ultimately thrive on Mars?

IMPROVISATION AND EXPLORATION

MASON PECK

Explorers of Mars will need many technical skills, not to mention training on how to operate technologies chosen to take them to the surface of that world and keep them alive there. Human exploration of Mars will require people adept at handling electrical and electronic systems, including generators, computers, communications systems and sensors; people who can repair life support systems and keep the air breathable, the water drinkable, and temperature comfortable; people who can survey sites and construct the habitats in which the explorers will live out the bulk of their lives.

But there is one technical skill that is more important than any other and that we can predict will be needed in abundance even without knowing all of the technical requirements of the mission: the ability to improvise.

Human settlement of Mars will likely be conducted in a way that astronauts of the last fifty years would find unfamiliar. From the pioneers of the American Mercury and Soviet Vostok programs to the scientists and sojourners on the International Space Station, astronauts follow carefully scripted procedures that safely plan their extravehicular activities and scientific investigations. And that's perfectly appropriate—for now. In the decades to come, we should expect our natural creativity, resourcefulness, and adventurousness to determine how we make the farther reaches of the cosmos our own, as our ancestors did when they left Africa tens of thousands of years ago. We will once again improvise solutions to problems we can't even yet imagine.

Certainly improvisation has played a role in space exploration. Some of us remember the Apollo 13 mission. Probably more of us remember the movie. At one point the three astronauts, Fred Haise, Jack Swigert, and Jim Lovell, moved into the Lunar Module to conserve power, saving what little the Command Module had left for Earth atmosphere reentry. But with three astronauts in that small space, carbon dioxide built up too quickly. They needed more lithium hydroxide canisters to remove it. They could not simply use the canisters from the Command Module, which no longer needed them. Those were square, and the Lunar Module used round ones. Engineers had to improvise an adapter to allow the square lithium hydroxide canisters from the Command Module to interface with the round environmental system on the Lunar Module. The fix required tape, cardboard, and plastic bags. That's all it took to put a square peg in a round hole. It was one of many unconventional solutions that ultimately saved the lives of the Apollo 13 astronauts.

> "I hope to obtain my destiny, which is to move mankind forward through technology, compassion, and an undying ability to perceive an enlightenment that does not yet exist. Forward thinkers are an integral part of human existence. Mars can bring us together as a species."
>
> —MARS ONE APPLICANT

And Apollo 13 is not the only example of improvisation in space. I've been involved in a number of so-called anomaly resolution efforts over the years—getting a spacecraft's deployable components, such as solar panels or antennas, unstuck from their cramped launch configuration by commanding unusual maneuvers; uploading some new software to allow a spacecraft with a broken appendage to continue to point at the Earth; and others I can't even mention because they're trade secrets.

I've seen unmanned spacecraft saved by all sorts of improvised solutions. In 1998, some of my colleagues at Hughes Space and Communications rescued the AsiaSat-3 spacecraft (later renamed PAS-22) after the launch vehicle had failed to insert it into a high enough orbit. Jerry Salvatore, Cesar Ocampo, and many others sent it to a geostationary orbit the short way—around the moon. Changing the plane of the orbit required less propellant that far away from Earth. It was the first commercial communications satellite to make that trip. It arrived in geosynchronous orbit with enough fuel to provide telecommunications services. I remember my colleague John Haskell calculating (by hand) carefully timed thruster commands to a spacecraft (one I won't name) to keep it from tipping over while we scrambled to plan an unanticipated orbit-raising burn. He got it right every time.

Anomaly means "something that's not the same." It's an unexpected turn of events. I suppose it's a euphemism, a nice neutral word in place of "screw-up," a more direct term that encourages us to point a finger at a scapegoat. A friend of mine once typed the letter *O* instead of the number *0* when he was sending a command to a spacecraft. It wasn't his fault, but that simple difference caused a severe screw-up—I mean an anomaly—that nearly destroyed the spacecraft. The littlest things can cause problems.

One reason these anomalies cause consternation is that we typically can't go fix spacecraft when they fail. NASA's shuttle missions to repair the Hubble Space Telescope are unusual exceptions, costing more than half a billion dollars each time. In most other cases, that cost isn't justified. It might be cheaper to launch a new spacecraft, and in any case a new spacecraft will last longer and include better features. So, satellite servicing has not caught on yet. Companies like ITT Exelis and ATK are working on new ways to achieve satellite servicing, but a viable business case will demand an innovative approach to reducing the cost. Even if satellite servicing does become a financially viable enterprise in the years to come, we're not likely to see routine service missions to Mars for a very long time.

So, we aerospace engineers plan ahead, trying to think of all the reasons why a space system might fail, and all this planning costs a lot of money. It also makes us quite risk averse. When we do encounter anomalies, the heart beats faster, reputations are called into question, and so on. We have learned to avoid them, taking the easy and conservative, if expensive, way out.

We incorporate redundancy and backup systems. We rehearse procedures that might be needed. The Jet Propulsion Laboratory (JPL) has a "Mars Yard," where a rover encounters all manner of obstacles and inhospitable conditions. Spacecraft operators must get the rover out of trouble, using only the existing commands at their disposal, honing their experience in using the hardware and the established operating procedures.

We do all that, but failures still happen, often because of something no one thought of. After all, if they had thought of it, the fix would already be in place. Despite all our risk aversion, our efforts at building fail-safe space systems, and rigorous verification of everything before launch, what makes spacecraft work today is that human element—the creativity and innovation that we see in successful problem-solvers from all walks of life.

"Opening a frontier for human exploration adds our priceless human abilities and traits to technology. We need technology as well as humans to colonize a new world."

—Andrea Keyser, Mars One applicant

Improvisation and in-flight repairs occurred often in the Russian space station program that ran from the 1970s to 1990s and was somewhat baked into mission thinking. For example, during the Soyuz TM-32 mission, astronauts fixed a video recorder with a soldering iron. True stories like this one from the Soviet era have inspired urban legends about space, including one you've probably heard: NASA is alleged to have spent millions to develop a space pen, while the Soviets gave their cosmonauts pencils. The truth is more prosaic: The Fisher Pen company had already invested its own money to develop a very robust design that also worked in space. So, NASA adopted it. And it's not that NASA didn't consider pencils. They did provide mechanical pencils to astronauts early on, but the danger of bits of carbon flaking off and drifting into electronics, as well as their flammability (remember the Apollo 1 fire), made them consider pens to be the better option. Still, this tale of simplicity and cleverness besting a government bureaucracy retains its appeal, and the story continues to circulate.

Here's a story of Russian technological improvisation with a greater likelihood of truth—although I must admit, I have not been able to verify it. Long ago, before NASA had ever docked the space shuttle to the Soviet space station Mir the Soviets were in a meeting with NASA space-station engineers discussing how such a maneuver would take place. NASA asked for vibration models of Mir, which are typically created as computer simulations that represent all the subtle structural behaviors of a spacecraft. The Soviets said that they did not have such models but could get some results soon. A short time later, they returned to NASA with beautiful, realistic charts showing the vibration response of Mir. NASA engineers were impressed that they could assemble a detailed computer model so quickly. The Soviets explained that they didn't model anything. They asked a cosmonaut to jump up and down at a specific frequency, and they simply measured the resulting vibrations.

Creativity and innovation is unavoidable in the context of an undertaking like Mars One. In fact, it will be central to making it work. Mars One colonists will certainly bring supplies from Earth, some unexpectedly useful and some unexpectedly useless. Some may be sent later. But what we can expect is that they will have to come up with many of their own ways to survive, and even thrive, on Mars. Ways to repair and improve their equipment, create an economy, new survival techniques, how to supplement their diets with minerals from Martian soil, ways to communicate with the folks at home, and even new ways to create art.

Many journalists seem to have a hard time distinguishing between science and technology. We hear "NASA scientists have built a new spacecraft . . ." No, they didn't. Engineers did that. Technology is creation, building, and innovation. Science is hypothesis, inquiry, and discovery. Both are distinctly human endeavors, traits that set us apart and make us proud to be who we are. But it's not science that will help a Mars One colonist put square filters into round holes. It's technology.

Technical problem-solving is what makes exploration possible. Informed by science, technology departs from merely understanding the nature of the world to create what has never been built before. Problem-solving skills of the type Mars One colonists will need come from making things, building them. These are the skills we associate today with the DIY spirit, whether it's from writing your own video game software, building your own car, or 3-D printing your own eyeglasses. Mars One candidates with the instinct to build, repair, create, and improve—they're the ones with the indispensable skills that will keep a colony alive and establish a permanent human presence on another planet.

A successful Mars One candidate isn't suspicious or intimidated by technology. In fact, he or she embraces opportunities to get physical with electronic circuits, mechanisms, and computers. Merely understanding abstractly how something works is a lot less valuable than being able to fix it and even build something better to replace it. Not all of these skills come from formal education. I can think of many engineering graduates I'd never let near a spacecraft. Instead, a Mars colonist is someone who is always learning new things and is eager to put that knowledge into practice.

One such engineer saved my father's life. About sixty years ago he was at the Naval Air Station in Pensacola, Florida, flying a Beechcraft military trainer known as the T-6 (or SNJ, in the Navy). After takeoff, his landing gear retracted, but the control handle for the gear stripped off; he wouldn't be able to lower the gear when it was time to land. These SNJ aircraft had been around for a long time, and a team of engineers were occupied full-time at Pensacola in those days to keep them running. One of the engineers got on the radio and walked him through a solution: remove the panel down by his feet (it required a Dzus key, but he used a coin), and loosen a nut that connects the landing gear to the control lever. With the gear free to move, my father rolled the aircraft back and forth as quickly as possible to shake the landing gear down into place. Another aircraft flew up underneath his and determined that the landing gear was "probably" locked. He landed easily, disappointing those fellow cadets who had hoped to see something more exciting that afternoon.

Such engineering know-how, preceded by years of engineers "designing out" failures, is key to the success of the Mars One colony. Hardware and software will be conceived, designed, tested, and implemented well in advance of the first boots on Mars. But every formal engineering discipline will also come into play during the daily lives of colonists on Mars: Civil, environmental, biomedical, mechanical, electrical, chemical, aerospace, and software engineering will all matter. So will the meta-discipline of systems engineering, which is concerned with the architecture and interfaces among parts of a large, complex system, such as a Mars habitat or an Earth-return vehicle. Here are a few reasons why they will come into play:

- We'll need civil engineers to implement the earthworks that will maintain a habitat's temperature and shield Mars One colonists from high-energy particles and ultraviolet radiation, particularly when the Martian regolith doesn't pile up the way we thought it would.
- Environmental engineers will figure out ways to keep the colonists supplied with water and air from the environment, when it turns out to be easier or harder than expected, all sustainably.
- Biomedical engineers will devise ways to create splints, eyeglasses, stents, prosthetics, casts, protective gear, and other devices that will become necessary as usage,

injuries, or unexpected medical conditions exhaust the supplies that came from Earth.

- Chemical engineers will know how to adapt smelting techniques to the local conditions, allowing the colonists to extract materials from the environment without needing to bring special-purpose equipment with them.
- Electrical engineers will repair, redesign, and create new circuitry to keep the colony functioning and make sure it doesn't fall too far behind the pace of technological innovation back on Earth.
- Aerospace engineers will track incoming spacecraft, build rockets for Earth return, and address unexpected problems that the Martian winds and dust storms cause.
- Software engineers will create new apps for the computing needs of the colonists, such as geolocation tools, games, and communications; and they will evolve the software that came to Mars originally, patching it with new code from Earth or of their own devising.

A lot of people with this spirit are already building their own spacecraft, and I can speak to that special group in particular. These days it's common for aerospace engineering programs in the United States to include a do-it-yourself spacecraft. College seniors launch their class projects. During my years at Cornell University, my students have built four small spacecraft, with three more on the way. But even some high school students are gaining space experience: Students at Thomas Jefferson High School for Science and Technology near Washington, DC, launched a grapefruit-size spacecraft that sent text messages back to Earth in 2013. In fact, this kind of small spacecraft is the most commonly launched type of satellite today.

These exciting trends in technology have us all thinking about the future of space and our place in it. Entrepreneurial space companies are building rockets, earth-observation satellites, satellite-servicing platforms, and asteroid-mining robots. In addition to Mars One, there are private plans to send people into orbit, to the moon, and to Mars in the coming decade, most of which will likely happen with the support of venture capital, not the more familiar government funding. At the same time, NASA is rediscovering its innovation roots, sponsoring the development of new technologies to push the boundaries of science and exploration. Add to the mix that individuals—members of the so-called maker movement—are passionately taking ownership of technology development, and we find that using 3-D printers and other cutting-edge additive-manufacturing technologies can accelerate the pace of putting hardware into orbit for commercial purposes.

In this new world, universities use spacecraft systems-engineering methods that are borrowed from the rapid development cycle of consumer electronics in order to reject decades-old principles that currently drive costs and schedules for spacecraft engineering. We can personalize exploration to produce research results with broad intellectual and societal impact that will change the face of our planet and beyond if we embrace the opportunities that small, agile space projects bring. Today's students have taken control of the means of producing space systems. Theirs will be a new generation with new design principles for personal spacecraft, using in-space resources to build exploration infrastructure across the solar system, and even benefitting from crowdsourcing to help them explore in ways we have never done before.

However, that isn't to say that improvising and drawing on local resources is without precedent, even without reaching all the way back to the great migration out of Africa. Plenty of examples of adaptation can be found in the history of the European settlement of North America. Evidence of how enterprising these settlers could be is on display at the Henry Mercer Museum in Doylestown, Pennsylvania. The museum houses an idiosyncratic collection of the tools of everyday life from the 1700s and 1800s. What I find remarkable in this collection is not only how cleverly people built their lives without the benefit of purpose-built tools and metals (even without nails, in many cases), but also their innovative use of what was at hand. Barrels, for example. Without the skills of a cooper, let alone the metal bands that hold the barrel staves in place, an early settler would cut down a large tree, section the trunk into barrel-size cylinders, and hollow them out. Instead of building a geared mechanism to raise and lower the bucket in a well, they would use a very long, lightweight tree, balanced on a fulcrum like a seesaw, to lever the bucket up and down in the well.

Clearly trees were plentiful, as was the space to build large equipment. The museum has a cider press built of two-foot-thick timbers to provide the mechanical stiffness necessary. The wood alone would make such a machine unaffordable today. What will the Mars colonists find in ready supply? Water and regolith. So, in the near term, earthenware, ceramics, and clay bricks make sense. Structures on Mars will likely resemble New Mexican pueblo dwellings or the early American pioneer sod huts more than they will contemporary visions of otherworldly, elegant structures of soaring aluminum and plastic. But I expect that the colonists will go through an accelerated sequence of human technological development in materials science—stone-age technologies such as mud bricks give way to ceramics, iron, bronze, and ultimately today's sophisticated materials as colonists innovate ways to produce material from the Martian regolith that we would not have thought of.

Mars One hopes to send the first colonists to Mars. It will be a one-way trip, which vastly simplifies the technology compared to what's required for a round trip (and makes it cheaper). Cynics assert that the thousands of people who have volunteered to be colonists must be out of their minds, that the trip is some sort of suicide mission. I disagree. (In the spirit of full disclosure, let me explain that I'm on the advisory board for Mars One. So, I would be expected to disagree with that view!) Mars One will send colonists to Mars not to die, but to live.

Here's the thing. I reject on its face that this trip is like the Kobayashi Maru scenario from *Star Trek II*, where failure is inevitable. One of the first projects I would take on after landing is to begin building the tools, the equipment, and even the economic structures to one day construct an Earth-return vehicle. Such a hope would not be irrational. In fact, it's easier to lift off from Mars and enter Mars' orbit than to do the same on Earth. I'm confident that the right group of engineers and hackers could pull this off over the course of a couple of decades, at least returning to cislunar space. And remember—they'll likely have internet access, probably with the help of laser communications. The folks back home can send them schematics, analyses, even test results, to help make that effort a success, which is more help than the American colonists ever had.

A space-hardware hacker might build something that takes advantage of in situ resources. Exquisite custom components will be unavailable or in short supply, like nails in the eighteenth century. The hacker is unlikely to find spools of carbon filament for high-tech composites, an unlimited supply of cryogenic valves, electric motors, circuit boards, or other high-precision special-purpose parts. Maybe a few items can be sent from Earth, on demand. I wouldn't count on it.

What we can count on is a ready supply of water, as much as the hacker wants. NASA has documented the abundance of water on Mars many times over. It's just beneath the surface in many places. Meteor impacts expose it fairly often. So, hydrogen and oxygen are no problem. They're available by electrolysis, which any kid can achieve on a small scale. So can graduate research assistants at Cornell University, by the way, where we have built a prototype rocket engine that runs on water.

Where would the hacker get propellant tanks for an Earth-return vehicle? I might suggest sand casting many identical equilateral triangles and welding them together to form a geodesic approximation of a sphere. We'll still need to ensure that enough aluminum is available. It would have to come from recycling bits of the lander, or other equipment. Alternatively, combining available aluminum powder with water produces H_2 gas and alumina (aluminum oxide, or Al_2O_3). That's a ceramic that has very nice properties, and it's strong enough for this application. The main advantage of using alumina is that Al_2O_3 is half aluminum by weight, thereby stretching the aluminum available to the hacker. A downside is that alumina tanks must be assembled by sintering, a kind of 3-D printing technique, and that's more time-consuming than melting down, casting, and welding aluminum triangles. But I suspect that something like a 3-D printer will be standard kit in homes and offices a decade from now, and a device that prints metal is the sort of tool that colonists would almost certainly bring.

Yet another solution would be to create watertight inflatable sacks. Cotton fiber has a very high strength-to-weight ratio. Worn-out t-shirts, lint from the clothes dryer in the Mars habitat, and some sort of rubberized liner might also do the trick. The result might even be lighter-than-metal tanks.

The water-electrolysis propulsion system in development at Cornell is simple, requiring no solenoid valves, cryogenic plumbing, or other subtleties. Its efficiency is lower than cryogenic H_2/O_2, but the simplicity can't be beat. Furthermore, the gaseous mixture is stoichiometrically ideal (i.e., there's just enough of each molecule to burn completely). As a gas, the combination is extremely well mixed, which eliminates the need for the combustion chamber to include an injector (and injectors are complicated and expensive to fabricate).

These examples are just some of the ideas currently under development: The ultimate test—and most likely the creation of even better ideas—will be in the hands of the engineers sent to Mars.

It will be essential for all involved in a one-way trip to Mars to proceed with a clear understanding of the risks: that is, the probability of success (or failure) and the consequences. But the real key to success to living on Mars will be the hacker ethos: Make do with what you have. Be creative. Your instinct for innovation will settle the solar system.

"We must know fear, we must understand fear, in order to make calculated risks. However, no amount of bravery can substitute for skill and knowledge."

—MARS ONE APPLICANT

***Dr. Mason Peck** has over twenty years' experience as an aerospace engineer and is currently an Associate Professor of Mechanical and Aerospace Engineering at Cornell University. He served as NASA's Chief Technologist from late 2011 through 2013. He directs the Space Systems Design Studio at Cornell, which is responsible for the Kicksat missions, the first crowdfunded spacecraft. His research focuses on technologies for sustainable exploration of the solar system and spacecraft architectures that democratize space. He is an advisor for the Mars One project.*

What medical skills will the crew need to survive and ultimately thrive on Mars?

MEDICAL SKILLS FOR AN INTERPLANETARY TRIP

The Hostile Environments of Space and the Planet Mars

THAIS RUSSOMANO

Imagine that you and your colleague have spent the last hour collecting small rock samples and are now on your way back to basecamp. You look out over the bleak and rugged terrain of Mars ahead of you, pockmarked with craters caused by the impact of meteors over thousands of years. You wipe your gloved hand over your helmet visor to remove its accumulated thin layer of red dust and look forward to being able to remove the restricting suit that is now feeling a little claustrophobic. Suddenly, you hear a cry of pain from somewhere behind you. Turning around, you see your colleague lying awkwardly on the ground, clutching at his leg with a foot that looks sickeningly splayed at the wrong angle. Now what? There are no ambulances on Mars, no emergency number to call, no paramedic to take control of the situation—there is only you. What will you do?

This scenario is potentially a very real situation that could be faced by any of the travelers on a Mars colony expedition. Traditional medical help will be unavailable, and communications with mission control on Earth will be subject to lengthy delays between transmitting messages and receiving any answers. Any event of this nature will need to be dealt with by the crew themselves, following set protocols and based on advanced training. This chapter will explore the scenario of a manned spaceflight to Mars and the hazards that will arise during the journey and while living on the planet. It will present some of the human physiological effects of remaining in a reduced-gravity environment and the resulting health implications, while also exploring the training that travelers will need in order to recognize health issues and deal with medical emergencies.

> "There is one Chinese old saying: It is not regretful to die in the evening if I can learn the truth in the morning. Life is short but the universe is ultimate. Human beings are so tiny compared to space. So if a tiny me could get closer to the truth of the universe, any cost is insignificant. If this journey does not work, at least I make my effort towards the truth and it will contribute to the final success."
>
> —Mars One applicant

On Earth, any significant trip you make, such as to visit relatives, go on vacation, or participate in a work meeting, begins with preparation. This is equally true of a journey to Mars. This preparation will include the acquisition of essential knowledge regarding changes that will occur in the human physiology during long exposure to weightlessness and the effects that different types of radiation will have on the human body on the way to the Red Planet.

The first step in acquiring this knowledge is to have a good understanding of the environment of the spacecraft in which you will be traveling and the Martian conditions that will await you upon your arrival.

All living organisms on Earth are subject to the influence of a gravitational force that has dictated the anatomy and physiology of terrestrial organisms for millions of years, including human beings. In general terms, gravitation is the force responsible for keeping the planets of the solar system in their orbits around the sun, the moon in its orbit around Earth, and the formation of tides. Every particle in the universe attracts every other particle with a force that is directly proportional to the product of their masses and inversely proportional to the square of the distance between them. Applying this to planet Earth, we call this force gravity. It is, therefore, the density of our planet, not just its size, that defines its mass and, consequently, the force that acts on every single object or being on our planet. Earth has a diameter of 12,775 kilometers (km) and a density of 5.52 grams per cubic centimeter (cm^3), resulting in a gravitational acceleration of 9.81 meters/second^2 at mean sea level, indicated by the symbol *g*. Mars is not only smaller than Earth, with a diameter of 6775 km, but its density of 3.94 grams/cm^3 is also less than Earth's density, which creates a gravitational force of approximately one-third that of Earth. This is called *hypogravity*, because we take Earth as our reference point—any celestial object that has a smaller mass, and consequently less gravitational force, is said to have a hypogravity environment. The opposite is also true—an object that has a bigger mass than Earth, and therefore greater gravitational force, has a *hypergravity* environment.

Earth's gravity has not only shaped our bodies, but is also responsible for the atmosphere we breathe. The atmospheric pressure on the Earth's surface is around 101.325 kilopascals, and the atmosphere is composed of 78 percent nitrogen and 21 percent oxygen, with traces of water vapor, carbon dioxide, and other gases. It is this atmosphere that protects us from the radiation of space and smaller falling meteorites. This atmosphere also moderates our planet's temperature by creating the greenhouse effect, a phenomenon in which molecules of the atmosphere act by capturing the thermal energy emitted from the ground, increasing the average temperature of Earth. Without this heat-retention effect, the average surface temperature of Earth would be –18 degrees Celsius, in contrast to the current +15 degrees Celsius, and it is likely that life would either not exist or exist in a different form.

Mars is such an attractive target for colonization because it has some important environmental characteristics that are similar enough to Earth's to make it potentially possible to sustain terrestrial life, with the help of advances in technology that have occurred throughout the last century. However, with a lower mass, the atmosphere around the Red Planet is thinner, with a mean pressure at surface level of 0.60 kilopascals (only 0.6 percent that of Earth). Therefore, Mars is not able to adequately protect living organisms from the radiation of space or from meteorites, nor can it create a greenhouse effect. Furthermore, the majority of its gas composition is toxic to humans, because it is basically formed of carbon dioxide (~95 percent). These challenges will have potential impacts on the health of explorers and will have to be monitored and mitigated.

Getting to Mars will involve a different set of environmental challenges, those posed by the spacecraft, which can also have potential health impacts. Space travelers will first endure a long voyage between Earth and Mars, during which they will experience microgravity and exposure to increased levels of radiation—the two major factors confronting their physiological well-being.

Advances in modern technology now mean that spacecraft have evolved greatly since the first flight of Vostok 1, piloted by Russian cosmonaut Yuri Gagarin in 1961. Currently, space vehicles maintain a standard sea-level cabin pressure and gas composition (20 percent oxygen and 80 percent nitrogen) that keeps cabin air temperature within the range of 18 degrees Celsius to 27 degrees Celsius, along with water-vapor pressure and carbon dioxide levels that are similar to those found on Earth. Nonetheless, the design requirements of the spacecraft will impose many restrictions on its crew, all of which may take a physical and psychological toll during the longer duration of a trip to Mars.

The size of the craft and the volume of storage area needed for equipment will determine its habitable space overall. Constant noise within the enclosed and confined space of the craft will require careful monitoring, and ear protectors will have to be worn if its volume exceeds safe decibel levels. The personal hygiene of all crew members will suffer; place four people together in a confined space for several months with minimal washing conditions and strong odors will soon become evident! These potentially harmful factors can have negative effects on the space travelers' sleep patterns, resulting in fatigue and raising stress levels that can lead to increased tensions among the crew.

A microgravity environment is one that imparts to an object a net gravitational acceleration that is extremely small compared with that created by the Earth on its surface. Microgravity conditions can be achieved in different ways, including Earth-based drop towers, parabolic aircraft flights, and aboard Earth-orbiting laboratories such as the International Space Station. As stated earlier, our anatomy and physiology have been dictated by Earth's gravitational force. As soon as this force is reduced or becomes absent, all body systems are affected and try to find ways to adapt, with the body as a whole finding a new way to structure itself and function.

The ear's vestibular system, responsible for our orientation and balance, is immediately affected by a lack of gravity, which can cause a disorder called *space motion sickness*, characterized by a loss in performance, nausea, vomiting, and a general lack of wellness. Astronauts are affected by nausea and vomiting to different degrees; for some, being in space has little effect, while others soon part company with the contents of their stomachs. In 1985, US Senator Jake Garn flew a seven-day mission as a payload specialist on the space shuttle *Discovery*. He went down in NASA history as the worst victim of space motion sickness when he vomited for much of his time in space. His legacy lies in the creation of an informal scale for space sickness, with "one Garn" being the highest possible level of vomiting and less severe cases measured as a fraction of a Garn.

Early on, the cardiovascular system also begins a process of adaptation to microgravity with a headward shift in the distribution of blood and body fluids, which affects blood pressure, heart rate, and cardiac output, and which also has pulmonary and neurological consequences. Astronauts have been known to suffer from other medical conditions, including variable degrees of anemia, immune and hormonal deficiencies, muscle atrophy and bone demineralization, sleep disturbances, and lack of appetite.

Not all of these physiological alterations are fully understood in terms of their establishment, evolution, and interactions, or how we might counteract them. The ideal plan would be the generation of artificial gravity through spacecraft centrifugation (rotating either a portion of a spacecraft or the entire vessel), but this is not a feasible option right now due to the large size that would be required of such a spacecraft.

Therefore, an important way of avoiding or decreasing the negative effects of microgravity on the well-being of a space traveler (and consequently on the mission as whole) is to take actions that prevent the evolution of these effects to diseases. This can be achieved through a wide spectrum of medical, technical, engineering, and organizational and logistics training and knowledge, all of which will be required to ensure adequate conditions while in space or when inhabiting another planet.

An example to illustrate the systematic approach required to prevent the progression of microgravity's effects is *bone demineralization*, the loss of bone mass that occurs over time during weightlessness exposure. This is a continuous process during space missions and can result in a severe case of osteoporosis—a disease more often found in the elderly here on Earth, which can lead to fragility of the bones and increased risk of fracture.

Osteoporosis in microgravity mainly affects the gravity-dependent bones, such as those of the lower limbs. If this maladaptation of the skeletal system to microgravity becomes too severe, the bones will be unable to sustain the body weight of an astronaut when he or she reaches the hypogravity environment of Mars. A daily program of exercise involving at least two hours of daily resistance and aerobic activities throughout the duration of the spaceflight will be vital to ensure that the new Mars inhabitants will be able to function well when faced with the planet's gravitational force. Assessments of cardiovascular fitness and muscle strength will have to form part of a regular schedule of medical monitoring, not only during the course of the interplanetary trip itself, but upon arrival at Mars and continuously throughout the duration of habitation.

Another major medical hazard that will face crew members during a trip to Mars will be the radiation exposure. Remember, on Earth we are shielded from much of the radiation coming from outer space by the atmosphere and magnetic field of our planet. However, we are still exposed to background levels of radiation, not to mention the commonplace minor exposure produced by some types of medical exams, such as x-rays, or from traveling onboard transcontinental flights, which fly at an altitude that means a thinner and less protective atmosphere above the plane. Additionally, natural sources of radiation can be found on Earth, such as radon, a colorless, odorless, and tasteless gas that is undetectable by human senses but gathers in places like basements and, when inhaled over time, is linked to lung cancer. Occasional devastating events have also occurred whereby radiation exposure threatens populations, such as after the meltdown at the nuclear power plant in Fukushima, Japan, in 2011.

In space, however, baseline levels of radiation are much higher than on Earth. Space travelers are continuously bombarded by cosmic and solar background radiation, increasing the risk of developing radiation-linked diseases. Astronauts in Earth's orbit still retain some protection from Earth's magnetic field, and although the spacecraft can also provide a degree of shielding from galactic cosmic rays, solar events, like coronal mass ejections, will cause large peaks in radiation. The space travelers will need to be alert to onboard monitoring equipment that will detect these radiation surges, and take actions to lessen their exposure by retreating to and remaining in a dedicated radiation shelter within the craft.

It has been possible to assess the potential levels of radiation over an interplanetary voyage following the advent of unmanned missions to Mars. It is estimated that the average dose received during a 180-day trip would be approximately three hundred millisieverts, the equivalent of twenty-four CAT scans. This would signify that a space traveler headed to Mars would be exposed to more than fifteen times the annual radiation limit for someone working at a nuclear power plant. This knowledge increases the significance and importance of appropriate mission and craft design, scheduling, and crew training, guided by a good understanding of the sources of space radiation and their deleterious effects on human health.

According to NASA, the health risks of astronauts and cosmonauts on the International Space Station are divided into four main categories: medical emergencies, radiation events, exposure to micrometeorite or debris, and system malfunction and failure. It is estimated that around three medical emergencies will happen during a fifteen-year period, and that there will be the need to evacuate a sick space traveler once in 5.5 years. Fortunately, however, the most common health issues are not dire medical emergency situations. These include eye injuries caused by floating particles in the microgravity environment; formation of urinary calculus (kidney stones) due to the calcium that escapes from the bones and deposits in the kidneys; dermatological infections, possibly linked to the limited hygiene facilities on board; and back pain as a result of the elongation of the spine that takes place when gravity is absent. Bearing in mind that a return to Earth for health care is impossible for travelers to Mars, health care systems aboard the spacecraft and on any Martian habitat must be designed with the capability to perform diagnosis and provide treatment for illnesses, especially in the preventative and early phases, as well as for acute injuries, such as our earlier example of the broken ankle. This space ambulatory care unit should be equipped to provide basic first aid as well as basic and advanced life support techniques, along with basic dental and mental health care. Ideally, the health care system would include equipment and devices that could support the performance of medical exams, such as pathological, imaging, and laboratory testing, as well as having a supply of medical consumables and pharmaceutical treatments.

It would also be very beneficial to the mission to have a fully trained medical doctor as part of the crew, particularly one with an emergency medicine specialization and experience in treating patients outside the confines of a hospital. The ideal candidate would be able to identify and handle physical traumas and neurological and cardiopulmonary emergencies. This will require the capacity to think and act quickly under stress with precision and determination. Medical training would preferably include challenges familiar to health care professionals who work in poor or remote areas of the globe, such as those who work for Doctors Without Borders (Médecins Sans Frontières), or in-the-field military doctors who take care of troops in war zones. Such doctors often operate in situations with poor communications, little or no outside supervision, and limited medical resources or assistance from other health professionals.

The ability to perform surgical procedures in space is constrained by the nature of the spacecraft microgravity environment. Some studies conducted during parabolic flights and involving animals in space missions have shown that minor surgery is feasible. However, even minor surgery has proved to be a complex affair. Restraint systems are needed for everything and everyone involved—the patient, doctor, and surgical tools—to prevent them from floating and moving around. Likewise, either a closed canopy system around the surgical site or an adequate suction arrangement is needed to prevent contaminating the atmosphere of the spacecraft with blood and surgical debris.

The reality is that surgical procedures, although not impossible, will need to be avoided whenever possible. NASA estimates that in long-term missions the indication for surgery will be largely related to acute conditions, such as appendicitis. A possible way of reducing future surgeries is through the prophylactic removal of the appendix and gallbladder of space travelers before their mission begins.

It will be necessary for all space travelers, regardless of their background, to undergo intensive training in the performance of basic first aid and emergency medicine procedures, as well as to learn about routine health monitoring, checkups, and how to use medical equipment. After all, the doctor may well be the one in need of medical assistance, and the time-lag delays in communication with Earth may limit remote guidance.

Providing clinical health care with remote assistance is called *telemedicine* and, even with communication delays, it will be a key player in the care of space travelers during interplanetary missions. Telemedicine has its origins in the US and Russian space programs of the 1960s and 1970s, and it has since grown into an emerging area of health assistance, research, and education on Earth, especially for patients in remote or impoverished regions. It can bring specialized care to people who would otherwise have to go without it.

However, here on Earth and during the low Earth orbit and lunar missions that have been conducted to date, it is possible to have real-time communication with medical specialists and advisors. Such back-and-forth contact will be extremely difficult once the voyagers travel more than a few light-seconds away from Earth. It will take up to twenty minutes for a signal sent from Earth to arrive at the planet Mars. This makes it essential for space travelers to be familiarized with and trained in how to handle medical informatics and the use of smart medical systems, which will need to be heavily incorporated into the health care system of an interplanetary mission. A smart high-tech medical support system will be designed to combine the means for acquiring and interpreting medical data, providing a diagnosis, and assisting in the decision-making process for the most appropriate treatment, based on the health care resources available.

On Mars, hazards like radiation exposure will be reduced, but other environmental dangers will present themselves. A 2005 NASA study summarized the risks that would be present for those living on the surface of the Red Planet, including Martian particulates from airborne dust and soil (affecting breathing), terrain hazards (accidents during exploration, especially incidents related to bone, muscle, and joint problems due to falling down), and environmental conditions (hypothermia that can affect neurocognitive responses). It will therefore be vital for a future Mars crew to be aware of and able to recognize these hazards and their negative health effects, while also being able to mitigate them with few resources and a lack of rapid communication with a medical support unit on Earth.

Habitation on the surface of the planet Mars will require that space travelers live in a pressurized and controlled environment. Nonetheless, extra-habitat activity will need to be undertaken on a daily basis. It is likely that the living modules on Mars will be pressurized to a similar level to that of sea level on Earth. The Martian atmospheric pressure is very low, so there is a risk that decompression sickness may occur due to moving back and forth from the habitat to open surface. Decompression sickness is a condition that can arise anytime someone moves from a high- to low-pressure environment. The problem arises because anyone venturing outside will need to don a spacesuit that, although pressurized, will still be set at a lower pressure level to enable a degree of working mobility for the person inside it (a fully pressurized suit would behave like a fully inflated balloon, and would thus offer considerable resistance to the motions of its wearer). This difference in pressures—the spacesuit having a higher pressure than the Mars environment but a lower pressure than the living modules—will require the suit to have a higher concentration of oxygen in order to avoid *hypoxia*, a condition in which the body receives an inadequate supply of oxygen. Hypoxia has a direct effect on the cardiovascular, pulmonary, and neurological systems, altering their function and impairing physical and mental performance.

Decompression sickness can also be life-threatening if left unresolved. The disease, more commonly known in diving as "the bends," is secondary to the precipitation of nitrogen bubbles that start circulating in the body and is most likely to affect large body joints like the shoulders and elbows, making them painful and reducing mobility. Decompression sickness can produce several worrying symptoms, including skin irritation; cardiopulmonary effects, such as chest pain and shortness of breath; and neurological implications, like altered sensation, amnesia, or vision abnormalities. The ideal treatment for the condition is repressurization with 100 percent oxygen in a special hyperbaric chamber. However, for each Mars inhabitant, early identification and avoidance of the disease will be the best solution.

The psychological aspects of long-duration space missions in individuals must be mentioned here because they are also a potential source of concern. Emotional issues related to isolation and confinement will quite naturally have an impact on the crew during the trip to Mars and also while living on the planet. Comments made by many astronauts who have stayed on the International Space Station have related to the comforting sensation of being able to see planet Earth so close; travelers to Mars, however, will slowly see Earth disappear from view, heightening their perceptions of separation, distance, and loneliness. Mental health issues, such as depression, anxiety, and stress, are likely to be just as relevant as the physical health of the Mars inhabitants and will need to be identified in the early stages to avoid progression and increased difficulty in management. A mixed-gender cross-cultural crew from different professional and personal backgrounds placed together in a confined space for twenty-four hours seven days a week will be subject to interpersonal issues no matter where they are placed in the universe, be it on Earth or Mars. The crew, therefore, will need to be exposed to psychosocial training and education regarding psychological signs and symptoms of behavioral issues and group interaction.

It is probable that many aspects of a journey to Mars will be automated within the craft, leaving the space travelers with available free time to be filled. This could lead to feelings of boredom and a sense of monotony, as well as heightened feelings of homesickness for family and friends. Measures will be required to occupy this time through the creation of stimulating activities, such as conducting meaningful experiments, using the computer or gaming, family contact—live initially and via recordings as the communication time lag increases—relevant and continuing study, and exercise programs. Unaddressed, the emotional impact of these feelings can be very negative for the psychology of the travelers, leading to depression, reduced social involvement within the group, and a lack of motivation for daily activities.

Fatigue can also become a problem in the enclosed environment of the spacecraft due to the constant noise of onboard equipment, irregular light patterns, and the absence of a normal circadian rhythm, all of which can alter the sleep-wake cycle and result in reduced sleeping time and/or poor-quality sleep. Eventually, this can lead to chronic fatigue, with its consequent disruption of mental and physical performance.

Living in a hostile environment, such as will be found in space and on the surface of Mars, will indisputably be a challenge and require a truly pioneering spirit from all those involved. However, humans have a long history of exploration and pushing the boundaries of their comfortable lives to reach into the unknown. Furthermore, technologies are rapidly evolving that will assist us with the goal of establishing a manned colony on Mars; solutions to those problems confronted will continue to be developed. The key element for the success and longevity of the mission will be based on a thorough and comprehensive program of training and knowledge acquisition, with components such as health maintenance and management, and accident prevention. Invaluable knowledge will be gained from the practical experience of life lived under conditions of reduced gravity in an atmosphere very different from our own. Much of this new understanding is likely to have spin-off benefits for us here on Earth, as has already proven true following the technological advances made as a result of the first fifty years of space exploration.

"I've always been drawn towards the unknown, the weird, and even the irrational. Aside from the obvious risks related to keeping the astronauts alive, and the uncertainties regarding how to provide them a life as long as possible, I see no disadvantages to participating to Mars One project."

—Mars One applicant

***Thais Russomano**, MD, MSc (Aerospace Medicine, Wright State University, USA), PhD (Space Physiology, King's College London, UK), founded and coordinates the Microgravity Centre-PUCRS, Brazil, internationally recognized for its space life sciences and eHealth research. She is a Full Professor at PUCRS and Visiting Senior Lecturer at King's College London. Thais holds international patents, has authored numerous articles and books, and acts as consultant and advisor for space projects. She is an elected member of aerospace academies and associations and is involved in space projects as Co-Founder, Corporate Director, and Chief Medical Officer of the USA-based International Space Medicine Consortium, Inc. (ISMC).*

What health and fitness skills will the crew need to survive and ultimately thrive on Mars?

HUMAN HEALTH AND PERFORMANCE FOR MARS MISSIONS

JAMIE R. GUINED

Decades of robotic exploration have taught us much of what we need to know about how to get equipment from the surface of Earth to the surface of Mars, but achieving the goal of putting "boots on Mars" will be a monumental undertaking fraught with risk and uncertainty. Of all the challenges that must be managed, mitigated, and overcome to enable mankind to turn science fiction into science fact, the "human-in-the-loop" is perhaps the most challenging of them all. In addition to the physiological, psychological, and engineering obstacles that must be considered and overcome to achieve this lofty goal, there are an equal number of accompanying ethical considerations that must also be taken into account.

The protection and maintenance of the health of the crew is one such consideration. According to NASA, the two greatest threats to human health for Mars missions are exposure to radiation and the chronic effects of long-term exposure to microgravity.[1] Although we've gained a significant amount of experience with extended stays in microgravity thanks to space station programs, the NASA Human Research Program has identified a number of human health and performance risks that it cannot adequately plan for, assessing our knowledge of them as either "Insufficient" due to a lack of data, or "Uncontrolled," indicating there is currently a gap in the knowledge that is required to adequately mitigate the risk relative to current NASA standards.[2] Consequently, the development of a Mars-centric research program about the effects of long-duration space travel must be undertaken in order to develop the capability and mission infrastructure that will be necessary to ensure human health.

But let's start with what we do know. The human body is a very efficient and resilient organism. There is a commonly held belief by the general public that living life in space is wholly detrimental to the human body and that as the amount of time spent living in the zero gravity of outer space increases, the functionality of the human body decreases. The truth is more complex, however, and depends on the context of the environment in which the human is expected to perform. The human body is very efficient in the process of adaptation. Astronauts and cosmonauts will eventually adapt to their environment and the demands placed upon their bodies, whether on Earth, Mars, or even the moon. Those who live aboard the International Space Station (ISS) adapt to living life and performing tasks in microgravity, which includes adaptive changes in human physiology that would be considered outside of "Earth-normal" physiological limits. Upon return to Earth they are considered to be somewhat *maladapted* to life under the influence of Earth's gravity, yet they are able to return to their preflight "baseline" level of conditioning within weeks to months after landing. Current (as of 2015) NASA standards of care for postflight reconditioning after a six-month ISS mission require two hours of exercise per day for the first forty-five days after landing.

In general, the physiological changes accompanying long-duration spaceflight include changes to the cardiovascular, musculoskeletal, neurovestibular, neurological, and pulmonary systems. Major physiological changes include bone demineralization; muscle atrophy with accompanied deficits in muscle strength, endurance, and cross-sectional area; neurovestibular impairments; neuromuscular insufficiencies; decrements to vision; and cardiovascular and pulmonary deconditioning.

The leap from long-duration ISS missions to a Mars surface mission is a giant one, with a plethora of unknowns and uncertainties as to the impacts to human health and performance. For instance, the current Mars mission architecture results in an increased exposure to radiation compared to that found in low Earth orbit, which could potentially exacerbate the chronic, degenerative problems mentioned above and ultimately affect the health of the crew and the success of the mission.

> "I am under no illusion that survival on Mars will take a lot of demanding work and that we may end up doing nothing else other than surviving. Equally the chance of premature death is high and although I have no death wish, I recognize its inevitability also and would want to make it mean something."
>
> —Mars One applicant

The effects of extended exposure (greater than six months) to zero gravity and hypogravity (gravity that is greater than zero but less than Earth-normal, as on the surface of the moon) remain to be seen. The standard duration for an ISS mission is six months, whereas Mars transits can extend up to ten months (depending on mission architecture), followed by the accrued—or unending—mission time spent on the surface of Mars. NASA ISS missions of longer duration, such as the one-year mission occurring during 2015 and 2016, have the potential to help researchers and mission planners better understand any relevant unknowns for future missions to Mars. Most likely the Mars One settlers will have to be their own guinea pigs, however, because the most valuable data for characterizing the effects of missions to Mars will come from the early Mars missions themselves.

The History of Exercise Countermeasures in Human Spaceflight

The use of in-flight exercise countermeasures in human spaceflight began during the NASA Apollo program of the late 1960s and early 1970s. During the Apollo missions, astronauts used a small variety of conditioning equipment, such as the Exer-Genie and Mark III, which were basically compact cable-resistance devices. The exercise equipment onboard Skylab included a stationary cycle, which was used by astronauts as their prime exercise device; an inertial wheel exercise device for upper-body exercise; and a Thornton treadmill for lower-body exercise.[3] The Skylab crews also used Lower Body Negative Pressure (LBNP) devices that simulated the effects of Earth's gravity on blood flow in the lower limbs, specifically to counteract the spaceflight-induced shift in body fluids toward the head. Crews on these early NASA missions exercised for a minimum of half an hour per day (Skylab 1) and a maximum of 1.5 hours per day (Skylab 4).

The Russian Space Agency (then Soviet Union) also conducted extensive research during this same time period that focused on exercise countermeasures aboard their Salyut and Mir space stations. Similar in concept to the LBNP device used by US crewmembers aboard Skylab, the cosmonauts employed the use of a so-called Chibis apparatus for the first time during the Salyut 4 mission. Other exercise countermeasure devices on the Salyut station included a treadmill, cycle ergometer, rotating chair (for vestibular research), and the Tonus, a device that provided an electrical stimulus to individual muscle groups.

The current in-flight exercise countermeasures program in use by the Russian Space Agency (aka Roscosmos) and NASA mandates that on-orbit crewmembers exercise 2.5 hours per day, six days per week as a countermeasure against the effects of microgravity. The primary goal of any good countermeasures program for human space exploration is to prevent a full adaptation to the space environment so as to minimize the detrimental effects of zero gravity–induced deconditioning upon return to Earth or another planetary body, like the moon or Mars.

After the NASA Skylab and Soviet Union Salyut programs, the inclusion of exercise countermeasures became a standard feature within human spaceflight mission architecture, with full integration aboard the NASA Space Transportation System, International Space Station, and the Russian Mir. Ongoing efforts in research and development for maintaining and optimizing crew health and performance through the use of exercise countermeasures are a priority among all spacefaring nations and programs.

Maintaining Human Health and Performance During Mars Transit

Maintaining human health and performance during transit to Mars will be no easy task. In addition to the challenges posed by living life aboard the confined and extreme environment of a long-duration spacecraft, establishing an effective exercise countermeasures program for the maintenance of crew health and human performance will add new levels of complexity and obstacles to overcome.

One obstacle will be the relatively small size of the spacecraft traveling to Mars. Crews on the International Space Station enjoy a multimodal exercise countermeasures program that utilizes a variety of exercise equipment. The ISS exercise countermeasures suite includes a cycle ergometer, treadmill, and a strength training device. However, this variety is only possible thanks to the sheer size and structure of the ISS.

It is essential that all Mars transit spacecraft be equipped with effective exercise equipment of high durability to prevent breakdowns during the journey. The countermeasures equipment should be designed for aerobic and anaerobic exercise and capable of producing both high- and low-intensity loads, and should allow for a range of individual exercise techniques for full-body strength training. The comprehensive countermeasures program for Mars missions should also include sensorimotor and vestibular adaptation exercises and training for the mitigation of performance decrements that result from the desensitization of the central nervous system brought about by extended duration in zero gravity.[4]

Astronauts in Mars transit must be prepared to respond to any number of in-flight anomalies or emergency contingencies that may require a high level of fitness, such as an emergency extravehicular activity (EVA) in zero gravity. A review of energy-expenditure data from NASA missions indicates that an astronaut can expend as many as 238 kilocalories per hour during an EVA, in part due to the increased effort required to compensate for suit pressurization.

"Being away from friends and family is a large sacrifice; in my opinion it's much larger than the physical endangerment component of the mission."
—MARS ONE APPLICANT

In addition to preparing astronauts for in-flight contingencies, mission planners should also carefully evaluate the tasks that will be performed in each phase of the mission. This critical analysis is necessary in order to better understand the physical and performance requirements for each mission task and the associated fitness and performance minimums that must be met in order to successfully complete a given task. This information can then be used to develop a comprehensive and targeted countermeasures program.

Ultimately, the use of artificial gravity—generated by large internal centrifuges or by spinning the entire spacecraft—is also something to consider for future mission-to-Mars countermeasures development. Future physiology and countermeasures research could possibly suggest that the use of an incremental gravity-loading countermeasure paradigm is a favorable and preferred technique for improving the body's response to the presence of gravity after prolonged periods in microgravity, such as during a Mars transit. As research efforts continue to advance in the use and effectiveness of artificial gravity, so, too, does our understanding of its potential as a countermeasure.

A successful mission does not depend solely on feats of engineering and rocketry but also on the human-in-the-loop. To ensure a safe and successful mission, planners must not overlook the importance of preflight physical fitness and general preparedness, the development and use of efficient and effective in-mission countermeasures, and comprehensive postflight reconditioning and rehabilitation.

The First Thirty Days on Mars: Readaptation to Gravity and Reconditioning

The first thirty days on Mars will perhaps be the most crucial for future Mars One settlers. After a lengthy Mars transit in zero gravity (six to ten months), the body will have undergone numerous adaptations outlined above that render it inefficient for surface life. Despite the inevitable deconditioning that will occur during Mars transit, crews will need to be able to immediately function at a potentially high level in the event of an emergency situation that could occur during descent to the surface of Mars. Likewise, once the crew has successfully landed, they will be faced with the necessity of beginning work on the establishment of the Mars One settlement, which will become the cradle of life on the Red Planet.

Upon landing, Mars One settlers will be forced to rely on themselves and on one another as they work under the stress of Mars gravity to safely perform tasks that may seem very simple to you and me sitting on Earth, but will be challenging to the settlers, starting with successfully exiting the lander spacecraft largely unassisted. Astronauts and cosmonauts returning to Earth after missions to the ISS are assisted during spacecraft egress by landing-support personnel who also assist with suit doffing. Unassisted, or minimally assisted, spacecraft egress and suit donning and doffing will require that astronauts and future Mars One settlers possess minimum levels of conditioning to complete these simple yet relatively demanding tasks. Of particular concern are the effects of long-duration spaceflight on the sensorimotor system and how the decrements observed in balance, locomotion, eye-hand coordination, perception, dynamic visual acuity, and gaze control will affect critical-task performance under the partial gravity of Mars. Such tasks include the ability to drive a rover and operate complex remote systems, such as telerobotic manipulators and other tools.

After the Mars One settlers have successfully exited the landing craft, they must immediately begin the critical work of constructing their basecamp. This condensed and demanding mission architecture will leave very little time for postflight rehabilitation or physical adaptation and acclimation. Therefore, it is reasonable to assume that mission designers and planners should give substantial thought to the preplanned workload and tasks to be carried out by Mars One settlers within the first thirty to forty-five days after landing. An incremental increase in task complexity and volume of work is recommended in order to provide sufficient time for readaptation and reconditioning to a gravitational environment while also minimizing the risk for injury or incapacitation that could result from a work schedule that is too aggressive and demanding. Mission designers should also try to factor in the use of such work-hardening protocols within the settlement's continued construction, surface exploration, and EHA (Environmental Health Administration) task plans. Implementing progressive work-hardening protocols will accomplish necessary settlement work while also providing a positive and incremental reconditioning benefit to the Mars One settlers, taking into account varying levels of operational readiness and fitness that will likely be present among the heterogeneous crew.

Establishing an Exercise Countermeasures Program for Permanent Mars Habitation

Just as on Earth, selection and implementation of a well-designed exercise program will be of paramount importance for maintaining optimal health and human performance among Mars settlers. Although our planet of residence within the solar system may change, our fundamental physiological needs as human beings remain the same, including our biological need for regular physical activity.

Everything begins with an understanding of and a proficiency in the fundamentals. Whether the subject at hand is mathematics, grammar, or physical movement, without a firm competency in the fundamentals, optimal advancement and proficiency will not occur. This is *especially* true for human movement. There is currently no method for re-creating Mars gravity (three-eighths that of Earth's gravity) for extended periods of time, which would be necessary for fully characterizing human locomotion in this alter-G environment. Each of the two methods used to simulate Mars gravity on Earth—parabolic flight and the use of specialized gravity off-loading devices—come with weaknesses that render them mostly ineffective for conducting meaningful research in Mars surface-task analysis, metabolic cost of EVA, and human locomotion. Similarly, these same simulations are of little benefit for training purposes. Upon landing on the Mars surface, the Mars One settlers will need to operate in an unfamiliar, novel environment. Just as with normal human-movement development and progression on Earth, including the development of the central nervous system, future Mars settlers can greatly benefit from a movement retraining program as a means of reorienting themselves with fundamental human-movement patterns that are not a part of daily existence in zero gravity (walking, crawling, stepping, climbing, carrying, jumping, and lifting), yet are a part of everyday life on a planetary surface. This ground-up approach to movement adaptation and learning is a precursor to more advanced and task-specific conditioning that focuses on developing the requisite level of readiness for construction and exploration tasks on Mars, and also assists in the "reawakening" of the brain's gross motor skills from its period of relative slumber during transit in zero gravity.

The value of integrated, effective exercise countermeasures programs for future Mars settlements cannot be understated; it is central for the survival of Mars colonists and should be considered a mission-critical element of all human space-exploration programs of extended duration and/or with planetary surface-mission architecture.

The needs of human physiological systems in space are no different from the needs of human physiological systems on Earth, with one caveat: The operational environment and its associated demands will determine the specific needs that must be met in order to maintain optimal levels of physiological functioning for each discrete physiological system. The human body is an amazingly resilient and adaptive organism that, when properly nourished and maintained through appropriate physical conditioning, skill training, and work-to-rest ratios, will find a way to achieve a new homeostatic balance in order to thrive and ensure the survival of the organism.

Jamie Guined *serves as Program Manager and a Commercial Scientist-Astronaut for Project PoSSUM (Polar Suborbital Science in the Upper Mesosphere), the world's first manned commercial suborbital research program, and as the Vice President of Business Operations for Integrated Spaceflight Services. Prior to joining Project PoSSUM and ISS, Jamie served as an Exercise Scientist with the University of Houston, supporting research at the NASA Johnson Space Center. Jamie holds a bachelor's degree in Exercise Science/Wellness, a master's degree in Health and Physical Education, and a Master of Business Administration, and is currently completing master's degrees in Health Science and Aeronautical Science.*

➤ Culture, Cohesion, and Compatibility

In his 2013 book, Mission to Mars, *Buzz Aldrin, the second man to walk on the moon, stated, "We need to have people up there who can communicate what it feels like, not just pilots and engineers." While the importance of the scientific and technical aspects of space missions cannot not be ignored, what is often neglected in the busy schedule of training a crew for flight is preparing its members for other, less explicit challenges: how to communicate, and how to work and interact together as a team and live effectively under very difficult circumstances. Two of the most important factors in Mars One's missions to Mars will be the training organized around effective interpersonal relationships and the selection of the most promising crews—not as this pertains to the individuals the crews are composed of, but rather the ability of crews to function well as teams. Careful crew selection, in this case a process that involves criteria that fall outside the current standards for space missions, and the ability to build not only a physical community but also an emotionally bound one, will determine whether we succeed or fail in this endeavor.*

Since the beginnings of human spaceflight there have been times when critical incidents relating to the psychological, behavioral, and interpersonal aspects of crew performance have jeopardized crew safety and mission success. A variety of physical stressors, such as temperature extremes, microgravity, solar and galactic cosmic radiation, and a lack of atmospheric pressure, characterize the spaceflight environment. While significant progress has been made in providing the engineering hardware and life support systems necessary for withstanding these stressors and sustaining human life in hostile spaceflight environments, long-duration missions also impose on crews significant demands for high-level team coordination and performance.[1]

In the history of mountaineering and polar expeditions there are many examples of crew breakdown, presumably caused by prolonged isolation and confinement, aggravated by severe environmental stressors: These stressors have resulted in individual psychological symptoms, decreased group cohesiveness, and have even jeopardized mission accomplishment. Furthermore, a report from the American Institute of Medicine indicates that differences in the cultural backgrounds of international crews are also an important issue facing long-duration space missions beyond Earth's orbit.[2] *For a mission to Mars to be safely undertaken with the international crew the Mars One Selection Committee members intend, any potential interpersonal difficulties must be addressed.*

Experience gained from long-duration spaceflights and space-analog environments has demonstrated that strong technical skills, excellent health, and the absence of psychiatric illness alone cannot ensure effective crew performance: Interpersonal and communications skills must also be stressed, as these essential elements help crew members work together effectively and lead to significant positive impacts on a team's performance.

In her captivating essay "A World Waiting to Be Born," Dr. Raye Kass, *an advisor to the Mars One project who has been involved in numerous space research projects in conjunction with the Canadian, Russian, and American space agencies, discusses the historical importance of incorporating skills beyond the technical, and how this must inform the process of how the crew of Mars One is selected, trained, and equipped. She also stresses the importance of how those who make the journey must understand the importance of the role emotional intelligence and interpersonal skills play not only in daily life on Mars but also during their journey to the Red Planet.*

This theme is further explored by Andy Tamas *in his informative and engaging essay, "Culture and Communication: Understanding and Managing Emotion and Conflict on the Mars One Mission." A development specialist in postconflict states with a focus on human resource management, intercultural team building, and communication, he focuses mainly on the problems that can be associated with cultural differences. He first explains the existence of different levels of culture and how the hidden dimensions of an individual's culture may impact his or her perceptions of other people. Tamas then explains how analyzing the intercultural relations competencies and team decision-making processes can help members of the Mars One crew manage their emotions and behaviors so that the crew can be more united and productive.*

The implications of age and aging on Mars' first settlers are explored in "Age and Aging on Mars." Compiled from interviews with thirteen of the one hundred remaining Mars One applicants, this piece details the perceived strengths that crew members who range in age could offer, as well as the challenges that could arise in a mixed age group. It documents the celebrations and losses crew members might feel while aging on Mars rather than on Earth, and how these two experiences may differ. The essay closes with a look at intergenerational teamwork and how this could play out in the new culture created on Mars.

Dr. Ronit Kark *wraps up this section with her provocative essay "Men Are from Mars, Women Are from Mars," which examines gender dynamics for the Mars One project. A tenured senior lecturer of organizational studies in the Department of Psychology at Bar-Ilan University in Israel, and a leader in the research field of gendered aspects of leadership, Dr. Kark explores the impact conventional notions of gender and gender role stereotypes will have on the candidate-selection process for Mars One, and the benefits of forming a gender-integrated crew. Finally, Dr. Kark explains how gender relations will need to be rethought when women and men from Earth land and live on Mars.*

—*The Editors*

What role do interpersonal and group skills play for the settlers on Mars?

A WORLD WAITING TO BE BORN

RAYE KASS

Operation Red Baron: Crew Set to Begin Journey to the Red Planet

By Catherine May (historian, biographer)
Space News International

FOR IMMEDIATE RELEASE: MARS ONE, YEAR 2026

Tomorrow morning at 1:47 a.m. UTC, a crew of four will blast off and take the first step toward a new chapter in human exploration. The first fully trained Mars One crew will be launched into the same Earth orbit as the crews of the Vostok 1 (1961), Apollo 11 (1969), Vostok 6 (1963), Challenger (1983), Endeavour (1994), and the Discovery (1998), to name a few. These expeditions allowed for many firsts in humans' journey to the stars: the first man in space, the first moonwalk, the first woman in space, the first untethered spacewalk, the first 3-person spacewalk, the oldest astronaut launched… the list could go on. However, notable in tomorrow's launch is a key defining situation. Unlike the men and women who came before them, the crew of Mars One will face a singular first: a permanent settlement on another planet.

The journey to Mars and the launch of the Mars One crew has been over 15 years in the making. Since the founding of Mars One in 2011, the public has been enthralled by this seemingly doomed expedition: a one-way trip to a planet that no human has ever stepped foot on.

Charting the Unknown: Behind the Curtain of Mars One

Imagine you are reading the above press release in the not-too-distant future: Implausible as it may seem, the world will soon be given this very chance to fulfill its dreams of space exploration and do something that has not been done since humans first stepped foot on the moon on July 20, 1969—set foot on a new world.

What is most intriguing about the unchartered territories of Mars is the unparalleled challenge and excitement of the human spirit willing to commit to a one-way destination, engage in the complex intensity of interpersonal relationships with no exit, and experience the painful and unending passage of time away from loved ones on Earth. What a mission to engage in!

What a vision of complexities for a human outpost on Mars—a planet that, although in our solar system, is virtually unchartered by human eyes! Apart from direct telescopic observations, all we know of the planet has been brought to us by our machine proxies in its orbit and on its surface. The Mars One voyage is more than just a mission; it's a journey into the unknown. In 2026, the first humans are scheduled to land on the Red Planet, where they will establish a human settlement. The Mars One organization will provide this unique opportunity, which will inspire generations to believe that all things are possible.

> "I am visionary and somewhat of a dreamer. I like the challenge of making the impossible into something possible."
>
> —GUSTAV ERIKSSON, MARS ONE APPLICANT

The men and women that will make this journey will need more than technological advances to see them through. Mars One crew members will be living in a community and relying on one another not only for survival but also for their mental well-being in the face of many unknowns. The unusual aspect of this Mars mission is that the crew eventually chosen will have *no known way of returning to Earth*. This is a one-way trip to the Red Planet.

This important and challenging dynamic is similar to those of settlers and explorers in earlier human history, such as the ancestors of the Polynesians, who pushed out across the Pacific to find scattered islands, or, hundreds of years later, Ernest Shackleton's famous Trans-Antarctic Expedition from 1914 to 1917. Shackleton's epic struggle to lead his twenty-eight men to safety after his ship got crushed in packed ice is a dramatic example of the challenges of the unknown. Voyages of this nature were dangerous and the risks could seldom be fully assessed, which is similar to the Mars One mission. Shackleton posted an advertisement to attract crew members and, despite being ominous in tone, the response to it was overwhelming.[1]

M. L. Barker, 1048 Chaplman Bldg.

MEN WANTED

for hazardous journey, small wages, bitter cold, long months of complete darkness, constant danger, safe return doubtful, honor and recognition in case of success.

Ernest Shackleton 4 Burlington st.

MEN - Neat-appearing young men of pleasing personality,

And although they did not choose to journey into the unknown, in August 2010, one hundred years after Shackleton led his crew toward the South Pole, thirty-three miners from Chile were trapped underground for sixty-nine days. They all survived, having managed to conquer the unknown with little to no preparation. Anyone who embarked on a voyage such as the Polynesians' or Shackleton's was also acutely aware that he or she might never return, and that it would be naive to suppose otherwise.

How can this be the case? What would possess people to leave behind all they know and journey into the unknown? Pondering this question has been at the forefront of the world's fascination with Mars One.

The Push into the Unknown: Strategy, Survival, and What We Know

Given the nature of the journey, one could assume that Shackleton's utmost concern was to find men who could fill and run the ship. But Shackleton was a man who understood the importance of choosing and nurturing a healthy team. An accomplished Antarctic explorer who had missed out on being the first to the South Pole (on an earlier expedition he had turned back ninety-seven miles from his goal to save lives—his own and those of his companions), he and his men were forced to fight for survival in an uncharted land. Because of the teamwork of Shackleton's crew, and Shackleton's own capacity as a leader, they survived against all odds and were rescued after twenty-two months.

Journey, Jesting, and Celebration: The Importance of Humor and Play

Shackleton was a master of group dynamics and psychology in unusual situations, and he was able to convey his own strength of spirit and optimism to his men, some of whom probably would have given up without his strong leadership. Shackleton was also adept at realizing what type of people he wanted along on his voyage: Navigation and exploration skills were important but not necessarily the defining factor. He chose musicians, storytellers, dancers—those willing to perform. The book *Leading at the Edge: Lessons from the Extraordinary Saga of Shackleton's Antarctic Expedition* describes how this helped the crew to maintain a high level of morale as well as ensure they had something to focus on other than the difficult journey ahead of them.

As Herbert M. Lefcourt notes in his book *Humor: The Psychology of Living Buoyantly*, to balance out the routine of their long journey into the unknown, Shackleton and the crew would gather each night to play games, sing, and celebrate birthdays and holidays. Crewman

Thomas Orde-Lees, who often considered these festivities a useless diversion, would later record one celebration as the happiest day of his life.

Shackleton insisted that everyone aboard participate in these events in an effort to prevent feelings of isolation and homesickness, which could demoralize the crew. And when the journey did become harrowing, with his ship and crew stuck in the ice, Shackleton proposed an activity that would raise the eyebrows of many a modern-day scientist: a group haircut. One by one the crew shaved one another's heads, with Shackleton volunteering to go first. This action served two purposes: to entertain the crew and to provide a symbol of their shared identity, strengthening their bond.

The aforementioned Chilean miners, trapped seven hundred meters underground, were all rescued alive. One miner, Ariel Ticona, was able to watch the birth of his daughter, Esperanza (which means "hope" in Spanish), via video and therefore keep abreast of the life he was fighting to return to. Another miner, Mario Sepulveda, explains how he felt it was his job to keep spirits up:

> Humor helped us. Sometimes it is easier to laugh than cry. I am a natural joker. When the others fell into despair, I would crack a joke . . . I have always been a clown. I knew it was important for me to keep joking around. If I stopped joking then I would be letting them down. But sometimes it was hard to joke. They talk about the tears of a clown. That was me . . . I had to keep that reputation going, for the sake of morale.[2]

Another miner made rescuers laugh by bringing up souvenir rocks with him as he was rescued and passing them out as gifts.

While no one is proposing that we install a juggler on the Mars One crew or stress the importance of a haircut during the mission, the above examples show that Shackleton and the Chilean miners understood something important: Even when in a crisis situation, groups need humor and play to help them function as a team. The application of this important lesson by the Mars One crew will be instrumental to ensuring that they make it to their destination.

The Decision of Discovery: Why Process Matters

Shackleton was also gifted at knowing when he needed to take a risk to move his expedition forward. He was willing to make difficult decisions, even when success was a long shot. He was able to maintain an optimistic outlook while being grounded in reality. The official photographer of the journey, Frank Hurley, had been developing and printing photos and putting them in an album. As Shackleton's ship, the *Endurance,* began sinking as its hull was crushed by surrounding ice, Hurley rushed to save his glass-plate negatives. Shackleton, tossing his gold watch and other possessions on the ice, ordered each man to save only two pounds of belongings. For Hurley, though, Shackleton relented, to preserve the record of the expedition for the future. Together they sorted through his negatives: When a plate was rejected, Hurley smashed it so there would be no second thoughts. He destroyed about four hundred and kept one hundred twenty, along with his album and movie film, storing everything in a can that he soldered shut. Hurley then threw away his movie camera.[3] When the ship eventually sank, the crew had to camp out on the ice in tents. Shackleton, trying to plan for the best in a worst-case scenario, eventually decided that he and the crew would have to head for the nearest land, nearly 350 miles away. In order to have any chance of success, he told the men they would have to be quite ruthless about taking with them only what was *absolutely essential*—which meant there would be no place for Mrs. Chippy, the carpenter's cat (a beloved member of the crew). Because the crew were entirely loyal to Shackleton and respected his judgment, this difficult decision, along with many others that had to be made, was carried out.[4]

Mine foreman Luis Urzúa, the last Chilean miner to be freed, said he kept the group together by putting everything to a democratic vote: "We were trying to find out what we could do and what we could not."[5] Miner Richard Villaroel said decisions were reached calmly and democratically: "If a decision was taken in which one person lost, most would still be winners."[6] Urzúa also said, "You just have to speak the truth and believe in democracy . . . Everything was voted on . . . We were thirty-three men, so sixteen plus one was a majority."[7]

A clear decision-making process ensures that decisions are made, even in a life or death crisis situation. In most cases, groups search for structure even when the decisions they make may have no outcome on their survival.

Why Work Works: Using Tasks as a Tool

Assigning meaningful and skill-based work was used as a tool by both Shackleton and the Chilean miners during their journeys into the unknown.

In *Shackleton's Way: Leadership Lessons from the Great Antarctic Explorer*, authors Margot Morrell and Stephanie Capparell describe how, at one point in the midst of a vicious gale, one of Shackleton's men said he wanted to die. Shackleton's response was to choose this man to replace the cook, who had temporarily collapsed. Shackleton later wrote, "The task of keeping the galley fire alight took his thoughts away from the chances of immediate dissolution. In fact, I found him a little later gravely concerned over the drying of a naturally not over-clean pair of socks, hung up close to our evening milk." The crew were assigned to contribute to the duties aboard the ship, including Shackleton, who would often lend a hand in the most menial of tasks and was likely to be the first one to step in and replace an injured or ill man. After the ship was lost to the ice and the crew's situation became more desperate, Shackleton assigned some of the routine duties on a rotational basis but, for other tasks, specifically selected team members based on personality. As a result, everyone's confidence and competence was boosted, note Morrell and Capparell, and social bonds among the crew were further strengthened.

For their part, the Chilean miners also felt it was important to remain engaged in meaningful work despite the probability they would never return to the Earth's surface.

It has been reported that the miners adopted different jobs based on their skills. For example, Yonni Barrios used his first-aid skills to act as the doctor for the group; Mario Sepulveda used his communication skills to present information in videos that were sent up to the outside world; Daniel Herrera was assigned the role of medical assistant, providing feedback about how efforts up above were affecting those down below. Sepulveda also stated, "We immediately organized ourselves into working groups to look for escape routes, keep the living area clean, and patrol the shaft for signs of rescue attempts."[8] Villaroel also supported this, stating, "We the mechanics were part of one group; we took care of the trucks. Other people organized the food; rationed it."[9]

The way a team works together is instrumental to what they can accomplish together. For the Mars One mission, the group's process during the journey may be of more importance than its scientific and technical tasks.

Okay, But So What?

But why are these things important? Why the need to ensure that the group we send to Mars is a team and not simply individuals sharing a common destination? If our technology can support a mission of this scope, surely the process of getting to Mars is merely the means to an end, with the destination, not the journey, as its emphasis.

We must realize where this crew will be going: a planet—in effect, an island universe. The advertisement Shackleton posted to attract crew members could also be used to describe the very environment in which the Mars One crew may find themselves.

Why Technology Is Not Enough

Mars One will provide an opportunity for scientists worldwide to work together and achieve what had previously seemed impossible: travel to Mars and its permanent settlement. Although significant progress has been made in providing the engineering hardware and life support systems necessary for withstanding the physical stresses of outer space and sustaining human life in a very hostile environment, is this enough?

Much work has been done by space agencies in the area of individual training and selection. However, experience gained from long-duration spaceflights and space-analog environments has demonstrated that strong technical skills, excellent health, and the absence of psychiatric illness cannot alone ensure effective crew performance. As long-duration spaceflight missions impose significant mental demands on crews, successful preparations for the mission to Mars must examine and address the difficulties that lie beyond the physical. Although we need to ensure that the crew who journey to Mars are adept at scientific and engineering thought processes (for obvious reasons), it would be disastrous to ignore skills that many pass over.

Currently, astronauts chosen for space missions generally fit the profile of "adults who take directions and follow rules like an exceptionally well-behaved child," as Mary Roach put it in her book *Packing for Mars: The Curious Science of Life in Space*. Today's missions have a specific set of training and preparation that lend themselves to a period of time with both a beginning and an end point, but this is not the case for Mars One. The mission of this crew will not end once the journey to Mars is over: It begins in earnest only when they have landed. And although the crew will train as best they can before they take off, the journey there will be their true training ground.

Once they land, the crew of Mars One will be faced with environments and situations that we on Earth have no concrete idea about. Shackleton's men journeyed into the unknown, with doubts as to their return. When the *Endurance* sank and the men began the most harrowing part of the expedition, they maintained a sense of optimism, despite the odds. What created their optimism? Most likely, the combination of their profound trust in Shackleton, their strong sense of community, and the confidence they had in their abilities as a team. The Mars One crew will need to develop these things as well: Once they are on Mars, they will be physically building a community in an unknown land, with challenges they will not be able to predict. But building a physical community is secondary to building an emotional community. It is essential for the crew of Mars One to understand this: Building a sense of community before they land will be imperative to their success.

Crisis in Space: Critical Crew Incidents

Since the beginnings of human spaceflight by the Russian and US space programs, there have been times when critical incidents relating to the psychological, behavioral, and interpersonal aspects of crew performance have jeopardized crew safety and mission success. Valentin Lebedev and Anatoly Berezovoy spent 211 days in space and often did not speak to each other. One wrote in his journal, "Humming to myself, I float through the station. Is it possible that someday I'll be back on earth among my loved ones and everything will be alright?"[10] On another mission, one cosmonaut cried out to mission control, "Fetch us back quickly. I can't work any longer with these zombies" (in reference to his fellow crew members).[11] Valery Ryumin was in space twice for six months aboard the Mir space station and wrote the following in his diary: "All the conditions necessary for murder are met if you shut two men in a cabin measuring 18 feet by 20 and leave them together for two months."[12] Vladimir Vasyutin, a cosmonaut

"I'm pretty easygoing, and get along with most personalities. That said, a wise person once told me that the best people to live with in tight quarters have 'a thick skin, a long fuse, and an optimistic outlook.' He was talking about crew for long sailing journeys, but I think his guidelines apply equally well to a Mars mission."

—Dr. Kimberly Binsted, Mars One applicant

on a 1985 space mission, had to be returned to Earth because he could not hold out any longer: He was reported to be depressed and overcome with a fear of death.[13] Protesting overwork, a crew on the US Skylab space station once mutinied and cut off all communication with Earth for twenty-four hours.[14] These examples highlight the need for more than just technical skills.

The Other Side of Being Smart

Buzz Aldrin once stated of those who chose to journey into space, "It is all very well to have technical skills. We have to learn to communicate." What did he mean? Surely we can be taught "to talk" much more easily than we can be taught to man a spaceship.

Teams are cauldrons of bubbling emotions. Virtually anyone who has worked on a team has noted the inescapable influence of the group's emotional underworld. And virtually anyone who has been a member of a task group has experienced the downward shift in the amount of work it produces when the emotional level of its members has not been taken care of.[15] In their book *Emotional Intelligence 2.0*, doctors Travis Bradberry and Jean Greaves state that, on average, we experience twenty-seven emotions each waking hour. With nearly seventeen waking hours in a day, we are likely to experience throughout the day about 459 emotional experiences. If one does the math, more than 3,200 emotions guide us through the week, and more than 170,000 each year. These are astonishing numbers that underscore the importance of building and nurturing a team's emotional intelligence, particularly as most of these emotions will occur during working hours.

According to scientists, the human body has two brains and two different kinds of intelligence: the rational brain and the emotional brain. These control how we think and how we act. Like conjoined twins, these brains are interconnected: When one gets upset, the other does, too. When feeling anxiety or distress, the emotional brain can take over the thinking brain and paralyze it. Although IQ is important, it is not everything. At best it contributes only 20 percent to the factors that determine life-success. Even among pools of high-IQ people, the most valued are those who can cooperate, collaborate, listen, support, empathize with others, and build consensus. It is not that IQ skills are irrelevant. On the contrary, they do matter, but mainly as "threshold capabilities."[16] What *can* make the difference? EQ, or emotional intelligence.

So what does this mean? What does EQ have to do with enabling the crew of Mars One to be more effective on their journey? Why is it so important?

The New Yardstick of Today

Emotional intelligence is the product of two sets of skills: personal and social competence. Central to emotional intelligence are self-awareness and self-regulation. In other words, EQ is being aware of one's emotions and knowing how to regulate and manage them. These two sets of competencies are directed both inward (toward self) and outward (toward managing others). Table 1 captures Daniel Goleman's four pillars of EQ: self-assessment, self-management, social awareness, and relationship management.

GOLEMAN'S FOUR PILLARS OF EMOTIONAL INTELLIGENCE

Emotional Intelligence Skill Areas	Emotional Awareness Abilities	Emotional Management Abilities
Personal Competence	Self-awareness of one's *own* emotions	Self-management (self-regulation) of one's *own* emotions
Social Competence	Social awareness of *others'* emotions	Relationship management of *others'* emotions

Table 1

Daniel Goleman wrote, "We're being judged by a new yardstick: not just by how smart we are, or by our training and expertise, but also by how well we handle ourselves and each other."[17] We are entering a new era in our world today, in which emotional intelligence—the ability to get along with people and make good decisions—is more important to life-success than the academic intelligence measured in IQ tests.

But how would a lack of EQ affect crews not working on this planet? Would there really be an impact?

In 1999, the Moscow Institute of Biomedical Problems (IBMP) was the site of a simulation of isolation, designed to help prepare for an eventual mission to Mars of considerably longer duration. Several agencies sponsored the simulation, including the Russian, Japanese, Canadian, and European space agencies. The subjects, in three crews of four, spent up to 240 days in a sealed chamber, similar to the duration withstood in the Mir Russian space station at the time.[18] The goal was to examine the psychosocial problems of living and working in an orbiting space station with an international, multicultural crew. Researchers examined problems that occurred during group interaction and tested countermeasures, such as specialized group-interaction training, to determine whether the acquisition of interpersonal skills would better equip astronauts and cosmonauts to deal with the psychosocial problems that arise from lack of teamwork during space missions. Crew members were asked to keep diaries with reflections about their time in the experiment, which would then be shared with researchers.

Although all of the crew members had high technical skills and backgrounds in space missions, their level of EQ essentials was not something that was taken into account when designing the experiment. This lack of consideration for the level of personal and social competencies needed by crew members led to critical incidents that, were they to happen aboard the mission to Mars, could lead to unmitigated catastrophe.

Catastrophe on Board: When EQ Fails

On New Year's Eve, one month after international crew members joined the Russian crew, an incident occurred when one crew member expressed jealousy of another crew member's ability to speak English.[19] This frustration erupted into a fistfight between the two, which in turn caused one of the crew members involved to be traumatized and isolate himself.

That same night another incident occurred: A female crew member was kissed against her will.[20] According to her diaries—and later confirmed during a debriefing interview—a crewman kissed her twice, intimately and against her will, which left her feeling violated and vulnerable. Interestingly, the diaries of other crew members, who were not present, did not mention this incident.

Following these incidents, one crew member, who was not directly involved in either incident, wanted to have the hatch closed between the two crews. The Russian commander informed mission control of this request. Mission control in turn suggested that the crew make the final decision. Most of the crew was not interested in having the hatch closed. But with the continued insistence of the one crew member, the rest of the crew acquiesced and they closed the hatch for three weeks. Once it was clear that a "visiting crew" would be entering the isolation chambers, the commander insisted that the hatch be opened. Interestingly, just before the hatch was reopened, the crew member who insisted on having the hatch closed quit the experiment and left the isolation chambers.

In many ways, the crew diaries revealed a high awareness and professionalism among the crew members in regard to what they were doing. In many ways, the diaries also revealed that many members possessed high *self-awareness* in regard to what they were doing but often lacked awareness of how their behaviors affected other crew members, and they received little feedback from colleagues about their behaviors or expressed attitudes. The diaries also indicated that the crew members didn't have an awareness that others might think differently from themselves. The crew's lack of ability to regulate their emotions and be sensitive to the emotions of others, coupled with their lack of relationship-management skills and cultural understanding, resulted in a disruption of congenial atmosphere. This lack of EQ essentials also effectively led to a breakdown in communications with not only the crew but also mission control. Were this to happen on Mars, people's lives and the mission would be in great danger.

The Action of EQ: How Can We Get There?

I have highlighted that a crew's level of EQ essentials helps in the development of a healthy and effective group community. As with the physical journey to Mars, the question is: How? What are the concrete action steps needed for developing and sustaining a team's emotional intelligence?

Table 2 (see next page) indicates what action steps can be employed by the crew members of Mars One to ensure that emotional functioning during their journey is as on track as the technical. These action steps help to foster individual and team awareness and regulation of emotions.

These important actions have proved highly beneficial as a part of the "tool box" belonging to individual group members here on Earth. For dealing with the highly charged, emotional atmosphere that is sure to sometimes be present during the crew's journey to Mars, they are imperative.

A Caveat About Emotional Intelligence

The skills shown in Table 2 develop only when we are open to learning and receiving feedback from others. They require one to practice, reflect, and mindfully develop the discipline of noticing. In other words, you have to work on EQ skills and be vigilant in their practice to reap any benefits, for they diminish in impact and value when they get no outdoor exercise.

CONCRETE ACTION STEPS TO BE TAKEN

Develop interpersonal understanding and perspective-taking.	Affirm, acknowledge, and reach out.	Create a trusting and safe environment.
Take time to understand a perspective that represents the opposite of your own.	Acknowledge contributions, work efforts, and the thoughtful interventions of others.	Exercise your discernment, integrity, and moral compass.
Check your understanding of the perspective of others before stating your own. When you do so, share how you are feeling.	Where possible, build on teammates' ideas, suggestions, and comments.	Allow for conflict to surface.
Respect differences in perspective.	Acknowledge feedback, even if you disagree with it.	Acknowledge differences and misunderstandings.
Periodically ask quiet members what they are thinking.	Acknowledge moments of caring.	Admit your mistakes and move on.
Check for an understanding of and commitment to the decisions being considered.	Provide emotional support when needed.	Avoid blaming, shaming, and/or gossiping about absent members.
Acknowledge your disagreement with any issues being discussed.	Listen, listen, listen.	Focus on problem-solving not blaming.

Table 2

Why the What of the Who Matters

Given the unique nature of the Mars One space mission, not only the current standards will be applied during the careful selection of its crew. An important aspect that will be considered is the willingness to build and maintain healthy relationships.

Buzz Aldrin once stated the following:

> No doubt about it. One-way Mars travelers will be 21st-century pilgrims, pioneering a new way of life. That will indeed take a special kind of person. Instead of the traditional pilot/scientist/engineer, Martian homesteaders will be selected more for their personalities . . . flexible, inventive, and determined in the face of unpredictability.[21]

The men and women who make this journey to Mars will need more than technological advances to see them through, because they will be living in community and relying on one another not only for survival but also for their mental well-being in the face of many unknowns. Although technology will get them there, teamwork will be what ensures they survive the journey.

According to the Mars One mission's website (Mars-One.com), the selection process for its crew will take into account the following attributes and requirements:

1. Gender identity (50 percent female, 50 percent male)
2. Age (minimum age eighteen)
3. Medical history
4. Nationality and language
5. Psychosocial factors

The ultimate goal of the selection process, slated to be completed by the end of 2016, will be to advance the most committed, creative, resilient, motivated, and knowledgeable applicants. Let's look at the psychosocial requirement for the selection process—after all, the Mars One crew will learn to get along because they have no other option. Right?

Unfortunately, it is not this simple: It is not enough to hope that crew members can eventually learn to both recognize and manage the inescapable influence of a group's emotional underworld. We must choose those individuals who exhibit the qualities that demonstrate their aptitude for interpersonal skills.

One key factor is *attitude*. It is the positional foundation upon which all else can be built. When one is on Mars, with no way of coming home, and questioning the decision to volunteer for the journey, what can help ground a crew member in his or her choice? A *deep sense of purpose* would help them to move forward, as one always needs to assess what motivates an individual to volunteer for such a journey. Another important aspect to consider is that of an applicant's willingness to *build and maintain healthy relationships*. This will be essential, as the Mars One crew will be living in community and will have to rely on each other for not only survival, but mental well-being in the face of many unknown factors for the rest of their lives.

The five key characteristics that need to be considered during the application and selection process for the crew of Mars One are outlined in Table 3.

Furthermore, one cannot stress enough the importance of an individual's capacity for *self-reflection*. Although I have listed five key characteristics that should be considered, paramount is the ability and openness to self-reflect. Without this essential foundation, the other characteristics cannot develop or be utilized to their full potential.

FIVE KEY CHARACTERISTICS THAT NEED TO BE CONSIDERED DURING THE MARS ONE APPLICATION PROCESS

Characteristic	Practical Applications (as seen and experienced)
Resiliency	• Maintains a robust thought process to persevere and stay productive • Sees the connection between the internal and external selves • Able to be at his or her best when things are at their worst • Maintains an indomitable spirit • Understands that the purpose of an experience may not be immediately clear but that a reason for it does exist • Asserts a can-do attitude
Adaptability	• Adjusts to situations and individuals while taking context into account • Knows own boundaries and how and when to extend them
Curiosity	• Asks questions to further understand things, not simply to get answers • Remains actively engaged in the journey toward knowing; not simply content with reaching an answer • Transfers knowledge to others; does not showcase what he or she knows and compare it to others
Ability to Trust	• Trusts in self and trusts in others • Trust is surrounded by good judgment • Knows when to start *mis*trusting • Possesses self-informed trust (trusts gut) • Reflects on previous experiences that help inform the exchange of trust
Creativity and Resourcefulness	• Able to apply these to the way an issue/problem/situation is approached • Not constrained by the way he or she was initially taught when looking for a solution • Uses humor as a creative resource in appropriate situations • Maintains a sense of play

Table 3

One Chapter Ends, but the Journey Begins

The human condition by its very nature is to push through crises: History is made when the impossible is made possible, and struggle is the impetus to change.

Despite the crises they will undoubtedly experience, the Mars One crew members will have to work together to create meaningful decision-making processes and ensure that the skills of all are being utilized to their fullest potential. It will also be important for them to celebrate the successes during a mission that is steeped in the unknown.

Like the explorers, pioneers, and adventurers before them, the crew of Mars One will hold the minds and hearts of those who watch them from Earth with bated breath when they launch into the sky on their one-way journey to the Red Planet. When the human spirit stops exploring, something has been lost.

The question is not whether we will make it to Mars, but whether we will be successful once we get there. And the Mars One mission begins not when the crew steps foot on Mars, but the moment the crew steps foot on the spacecraft.

Prof. Dr. Raye Kass, *Professor of Applied Human Sciences at Concordia University in Montreal, Canada, currently spearheads group theory courses in the graduate-level program.*

Known for her contemporary, cutting-edge, and timely research style and results, Dr. Kass has been highlighted frequently by both national and international press agencies for her space sciences and group theory research. Dr. Kass has also been invited to be involved in numerous space research projects in conjunction with the Canadian Space Agency and NASA, including the Psychological Experiment/Training Programme for the CAPSULS Mission held in Canada, the SFINCCS mission held in Russia, and the NSBRI (National Space Biomedical Research Institute) Ground-Based Research Project with the NASA Ames Research Center in the United States.

Dr. Kass is the author of Theories of Small Group Development, *as well as the coauthor of three other books on group theory. She has also published more than eighty scholarly works, including refereed chapters in a book, book reviews, refereed journals, non-print publications, unpublished manuscripts, reports and studies, research proposals, and refereed conference presentations.*

What role does cultural background play for the settlers on Mars?

CULTURE AND COMMUNICATION

Understanding and Managing Emotion and Conflict on the Mars One Mission

ANDY TAMAS

The Mars One team is likely to have members from a wide variety of cultural origins. The differences in their early childhood programming can be both positive and negative: They can combine in creative ways to make something new and wonderful and also contribute to interpersonal conflict. Although the positive side of intercultural relations can be more interesting to explore than the negative, this piece will focus mainly on the problems that can be associated with cultural differences. The emotions and conflicts associated with interpersonal communication, particularly where differences in culture, values, and cognitive processes are a factor, need to be understood and managed if the Mars One mission is to succeed. This applies not only to those candidates chosen for the mission but also to all of us on Spaceship Earth.

"The hard skills are the soft skills," said a seasoned software engineer at a team-building workshop I was facilitating in the high-tech sector some years ago. The members of the team were having difficulty combining their varied talents—and diverse personalities—to create the new products they knew they had the capacity to develop. The engineer had hit the nail on the head: Each member of the team had tremendous technical talent, but they needed something else—soft skills—to combine their diverse capabilities and form a unified and harmonious whole to facilitate the breakthrough results they desired. Each member of the team needed to better understand and manage emotion communication and conflict and how to handle themselves in critical high-pressure interactions with people who thought, felt, and behaved differently from themselves—factors that were not part of their engineering training. Mars One calls for these soft skills, in spades. Fortunately, they are relatively easy to identify and learn how to manage.

A good place to begin is to reiterate the well-known fact that most of what is exchanged in interpersonal communication is *not* in the words. Some analysts say that over 80 percent is in other parts of the interaction—the context, tone of voice, nonverbal communication, and so forth. So how does the perception system work to receive and analyze this much larger component of communication? Understanding this wonderfully complex dimension of ordinary, everyday life requires some insight into how the cognitive system operates.

Filtered and Distorted Perceptions

The senses are able to receive and analyze far more data than the bits that surface in our conscious awareness—much of the rest is processed in the background. Actually, the perception system is a selective distorter of sensory input that usually admits to our consciousness only those images that find a match with our previous experience, with pre-existing cognitive categories. We usually can perceive only that which is similar to what we have previously learned to see and hear, and input from unfamiliar sources is often either invisible or distorted to match known categories or images.

One day, in the Canadian arctic, I went out hunting for caribou with an Inuk friend. We climbed up a small rise and looked over the treeless landscape to see if there were any caribou. My friend said, "We're in luck—there are about a dozen out there." Yet I could see only rocks of various size and color and a few small ponds and low bushes, and the absence of trees meant I had no visual scale to help me judge size or distance—it was quite disorienting. I looked through the binoculars he gave me and still I could see only rocks of various colors—until one of those big rocks changed color, from dark gray to light gray. The rock had moved—and had legs and antlers. It was a caribou.

> "Because I was born in Japan and spent my formative years there, it's been easy for me throughout my life to engage, learn about, and understand different cultures. I've even visited the interior of Mexico on a sketch-journal adventure, visited New York many times, and spent a whole summer on the Alaskan peninsula living in a tent working in salmon fisheries, as well as almost a whole year on a fishing trawler based in Unalaska. Now that's culture!"
>
> —Mars One applicant

I had never before seen a caribou in the wild, and my perception system was playing tricks on me: The visual data from the caribou had been distorted to look like something familiar—a large rock. As soon as I saw the first one, I could see the others; I had learned what to look for. What we perceive is filtered and distorted by our cognitive system to conform to what we are accustomed to seeing, yet it may not be what is actually there.

The lesson in this for the members of the Mars One team is to be aware of the working of this system and not put blind faith into what their senses are telling them. This can be a challenge, though, because the way we have learned to see and understand the world is the foundation of our cognitive framework, which is the basis of our identity and personal culture—elements that are difficult for us to perceive, let alone change.

Diversity, Culture, Rules, Emotions, and Human Relations

The diverse members of the Mars One team will need to know a lot about culture—beliefs, emotions, and human relations; the soft skills noted earlier. Anthropologists Geert Hofstede and Edward T. Hall spent decades studying these issues. Hofstede called culture the "software of the mind" that determines how people and organizations function. Like most good software, it runs in the background; it is largely invisible when it is working as expected. However, this software operates differently among peoples of different cultures. These differences can contribute to problems when they interact.

> "My time at my previous job brought me together with many people from diverse backgrounds. There were observable differences in our food, dress, and word choices based on our cultures but, all in all, there were no problems."
> —Mars One applicant

Edward Hall described a group's culture as a complex set of largely unconsciously held rules that are learned early in childhood but soon disappear from conscious awareness; like a hidden submarine, they drive from the depths. Most of the rules we learn are "normal" or taken for granted as the way things ought to be. As a result, one's own culture is largely invisible to oneself: It is the uniquely formed lens through which we, and others like us, make sense of the world.

These protocols are not all of the same order or degree of visibility. Hall used the concept of an "iceberg of culture" (see Figure 1, page 81) to identify three layers of learned rules: The small visible part above "sea level" he called the *technical* layer of rules, which can be relatively easily learned by outsiders. Hall placed language, art forms, traditional dances, and other similar components in this layer, and said that changes or infractions of the rules associated with this dimension usually do not elicit strong emotional responses from members of the group.

For instance, I speak French fairly well but often make errors in gender, using the male form for female terms and vice versa. When I give workshops in French, these errors are usually overlooked by participants and don't cause them to get upset—they attribute these to my lack of fluency in French, an excusable category of error or rule infraction.

Hall next defined a *formal* set of rules as a relatively small layer that is partially above and below the waterline—some are visible, others are not. He placed rules associated with table manners, interpersonal-space norms, and other similar social patterns in this layer. Infractions of rules at this level can elicit relatively strong emotional responses that could negatively affect interpersonal relationships. Bad table manners, for example, can have a negative impact on the chances of receiving a promotion to higher management for which one is technically qualified.

The far larger set of rules he called the *informal* layer, which is completely below sea level and almost completely invisible even to members of the culture. Infractions of these rules usually elicit intense emotion and prompt strong negative reactions that can often have serious impacts on interpersonal relations.

I have made a number of errors in this dimension during my organizational-development work with people of various cultures. In one case, I mistakenly assumed that a senior-level Middle-Eastern manager who said he wanted to adopt modern open and participatory management processes was ready to freely discuss some of the problems he was struggling with, and raised the issue in a management team meeting without discussing it with him privately beforehand. It was a stupid, interculturally incompetent move on my part: His status was threatened by my initiative and he lost face in front of his group. I lost that job and was soon on a plane heading back home. I doubt that he could identify precisely which cultural protocols I had violated—all he knew was that he did not want me around any longer: I was toast. His technical-level desire to adopt Western participatory-management processes collided with his deeper-level protocols associated with identity, status, and position in the social and organizational hierarchy. I was not sensitive enough to know this might be the case and blundered badly on that particular job, even though I had been doing intercultural relations training and organizational development work for decades before that.

> "I want to inspire not just others in the way of universal harmony and knowledge, but continue to inspire myself and show that nothing in life is ever unreachable or unattainable. Some people wonder how their lives could be different if they could just start over again knowing what they know now. This opportunity is just that. It provides individuals from different backgrounds and ways of life the opportunity to come together and start something new and exciting."
>
> —Mars One applicant

Hall's three layers of culturally learned rules are illustrated in Figure 1—the deeper-level protocols are almost completely unconsciously held, yet exert a great influence on our emotions and related behaviors. Problems with interpersonal relations can be explained using this iceberg analogy. When different icebergs of culture come together, as they would in the close quarters and long-term relationships among Mars One crew members, there can be collisions at the deeper levels—often unintentional infractions of invisibly held rules. These violations can often produce the intense emotions and subsequent disruptive behaviors that negatively affect the quality of interpersonal relationships and group cohesion and harmony. Consequently, they could also lead to mission failure for the crew of Mars One.

Because the infractions at the informal level take place in the depths—where that hidden yet functioning submarine propels each of us through life—they are not often evident to the actors themselves and

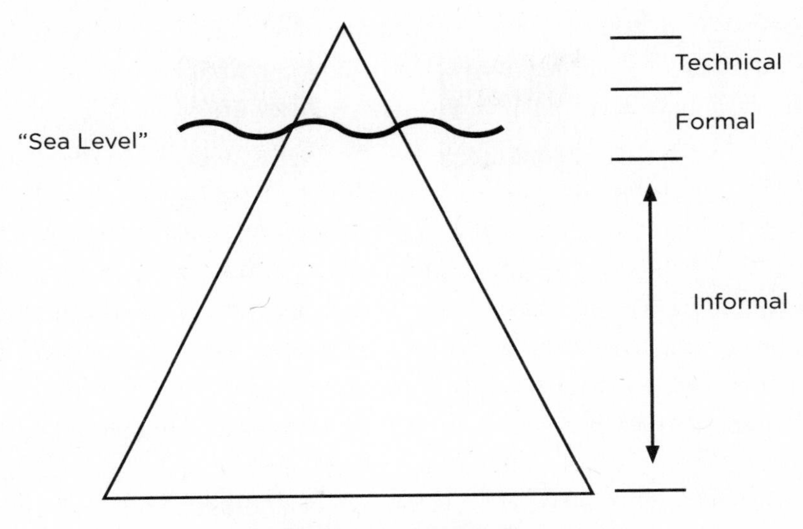

Figure 1: Levels of Culture
Adapted from Edward T. Hall's *Beyond Culture* (1976)

therefore not readily subject to the examination and revision required to maintain harmony in the group.

The hidden dimension of the rules or protocols of one's culture forms the basis of personality and identity. Irreversible changes in one's context (such as a major physical, social, or economic disruption) that affect this deepest dimension can cause intense discomfort and elicit powerfully disruptive behaviors, including violence, depression, and even suicide. Peter Marris, in his book *Loss and Change*, described the impact as similar to the feelings of bereavement that accompany the loss of a loved one. People will go to great lengths to avoid that experience, which is linked to identity-maintenance behaviors and the normal human need to maintain the integrity of one's personal structure of meaning.

Remember, it is mostly the learned cognitive patterns of our culture that provide the lens through which we perceive and understand the world: It was the hidden yet active mechanism that tricked me into believing that a caribou was a rock. This system can operate in another subtle way as well. Research on interpersonal relations indicates that people can have a sense of whether or not an individual is being completely honest and transparent—whether they are what is known as "self-congruent"—by picking up and analyzing his or her small behavioral cues using the largely subconscious invisible-analysis system described earlier. And although this perception system can work differently for each of us—being primarily a product of an individual's personal culture—the result of this analysis surfaces as a feeling, or gut-level hunch, that can be positive or negative. Mars One team members need to be conscious of these processes and the emotions they can generate, and learn to self-manage their feelings and human relations behaviors so they can maintain unity and harmony in diversity as they go through their mission. This is the core of emotional intelligence and also of intercultural-relations competencies.

Measuring Intercultural Relations Competencies

Some people are better at managing intercultural relations than others. Fortunately, the required competencies can be measured. One such instrument is the Multicultural Personality Questionnaire (MPQ), developed by Karen van Oudenhoven Vander Zee and Jan Pieter van Oudenhoven, respected cross-cultural psychologists in the Netherlands.

The MPQ is an online personality-assessment questionnaire that can be used to predict how easily people are likely to adjust to other cultures and come to feel comfortable with them. It measures the five capabilities that are present to varying degrees in virtually everyone: cultural empathy, open-mindedness, social initiative, emotional stability, and flexibility. Descriptions of each are as follows:

1. **Cultural empathy.** This scale assesses the capacity to identify with the feelings, thoughts, and behavior of people from different cultural backgrounds. To function effectively with people of other cultures, it is important to

acquire some understanding of those cultures, and cultural empathy seems important to "reading" other cultures.

People who score high on this scale are able to identify with the feelings, thoughts, and behaviors of people that are part of different cultures. Those with low scores usually have more difficulty.

2. **Open-mindedness.** This scale assesses people's capacity to be open and unprejudiced when encountering people outside of their own cultural group and who may have different values and norms. This ability, just like cultural empathy, seems vital to understanding the rules and values held by other cultures and to interacting with them in an effective manner.

   People who score high on open-mindedness have an open and unprejudiced attitude toward other groups with different cultural values and norms and are open to new ideas. People who score low are characterized by a predisposed attitude and a tendency to judge and stereotype other groups.

3. **Social Initiative.** This scale denotes the extent to which people approach social situations actively and take initiative to interact with others. It determines the degree to which they interact easily and make friends with people from different cultures.

   People who score high on this scale tend to be outgoing when interacting socially with another culture. People who score low are less inclined to take initiative: They tend to be rather reserved and stay in the background during social interactions.

4. **Emotional stability.** This scale assesses the degree to which a person remains calm in stressful situations. When working in a new cultural context it is important to be able to cope well with psychological and emotional discomfort.

A variety of factors (e.g., political systems, procedures, lack of means and resources, impediments) may cause things in other cultures to work differently than they do in one's own culture. When things do not go the way they would in one's own culture, it can lead to frustration, tension, fear, social detachment, performance problems, and interpersonal conflicts.

People who score high on this scale tend to remain calm in stressful situations, while people who score low exhibit strong emotional reactions to stress.

5. **Flexibility.** This scale is associated with people's ability to adjust their behavior to new and unknown situations. When working in a new cultural context, it is important to be able to change strategies, because customary and trusted ways of doing things will not always work.

   People who score high on flexibility perceive new and unknown situations as a challenge. They are able to change behavioral patterns in response to unexpected or constrained circumstances in another culture.[1] People who score low are quicker to see new and unknown situations as a threat, and they tend to stick to trusted behavioral patterns. Consequently, they are less able to adjust their behaviors in reaction to unexpected or constrained circumstances in another culture.

Because—as was mentioned earlier—one's culture is a product of the rules learned early in life that have largely disappeared from conscious awareness, it is likely that each member of the Mars One team will have his or her own individual culture with its largely invisible patterns of perception and emotional responses. The five capabilities measured by the MPQ should be taken into account when preparing for the mission.

All interpersonal encounters are, to some extent, intercultural encounters. Even if the team members are from the same society, it is likely there will be differences in cultural patterns. Identical twins raised in the same home can show somewhat different personality traits, variations that can contribute to interpersonal-relations challenges.

Research has shown that cultural differences between members of similar societies (such as Americans and Canadians, for example) are more difficult to identify and manage than between members of obviously dissimilar societies. In the latter there is an expectation that there will be different patterns and people are ready to compensate for them. In the former, however, miscommunication tends to be less attributed to cultural differences than to personal failings or inadequacies in the individuals involved. Rather than being ready to understand and compensate for miscommunication problems, people who engage in difficult encounters with members of similar cultures can become impatient with one another, and their frustration can contribute to destructive conflict in the relationship.

Conflict and Human Relations

Conflict is a normal and necessary part of human behavior and can be destructive or constructive. It also plays a major role in how teams function. There is likely to be some conflict among the Mars One team members, so their preparations should include strategies for conflict management.

Teams need to foster and properly manage constructive conflict if they hope to survive in rapidly changing environments. Too much harmony can be dangerous—it can lead to lemminglike behaviors that take a whole group over a cliff. When conflict is suppressed or is unable to be used in a positive way, it often manifests itself in a destructive manner.

The two types of conflict are associated with two sets of attitudes or human relations characteristics, as shown in Figure 2.

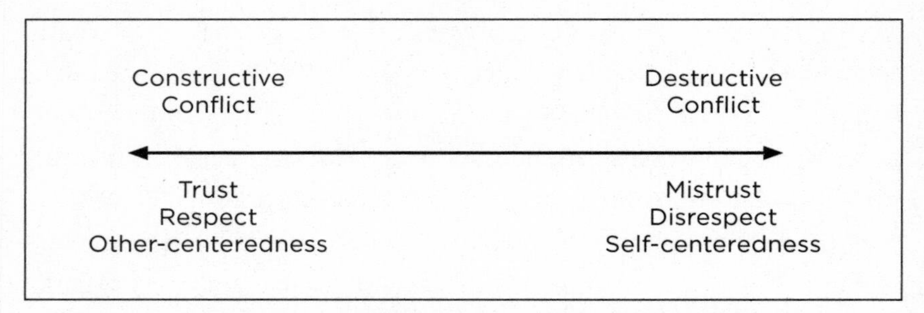

Figure 2: Conflict and Quality of Human Relations

Tamas, A. *Managing Diversity and Change* Training Workbook (2001).

Destructive Conflict

When people think of conflict they often think of the destructive end of the continuum, with high levels of negative emotion, arbitrary and uncaring use of power, loud and disrespectful verbal exchanges, and the possibility or reality of physical violence. This kind of conflict can foster paralysis, in that it makes it impossible for people to work together, and can result in explosive and uncontrolled change that may cause significant harm to those involved.

Constructive Conflict

When conflict is constructive, the parties express differing opinions in a way that fosters a positive change in a group. The team creates conditions in which individuals freely and frankly express their opinions, question one another's sets of assumptions, and build on one another's contributions to discover the path to a more positive future. This orientation to teamwork is a core element in fostering creativity and growth in groups.

Managing Conflict

A key to managing conflict is to have agreement on communication strategies—a common understanding of how the team is to function. There must be what is known as a "common grammar," or rules of engagement for discussion. Members of a group that want to manage conflict need the following:

1. To agree on the underlying protocols for communication (i.e., a common grammar). Robert's Rules of Order provide a rather primitive set of protocols. More advanced examples are the *process* elements described in the next section on team decision-making.
2. To know what preconditions must be met for the protocols to operate.
3. To understand that the rules may seem artificial but are nevertheless useful. There is no reason to keep doing things as they have been done in the past: Like culture, one's strategies are largely transparent, but, because they have been learned, they can be changed.
4. To know the procedures to change these rules of engagement. An example is a decision to switch from a consensus-based decision process to a majority-rule system once it becomes clear that consensus cannot be achieved. A further shift can take place when a member who "lost" a vote refuses to abide by the decision of the majority—the hierarchy can then enter the picture and raw power can be used to impose a choice. This illustrates three kinds of rule-sets—consensus, majority rule, and raw power—in declining order of desirability.

In their article on conflict styles in *Interpersonal Conflict,* Hocker and Wilmot say that people who have developed this critical and self-aware perspective and sensitivity to procedure are sometimes referred to as "rhetorically sensitive," meaning they can do the following:

1. Maintain a flexible set of roles and communication protocols so they are able to adapt as required;
2. Withstand the ambiguity of constant adaptation and develop skills in dealing with different audiences;
3. Monitor their discourse to make it purposive rather than simply expressive—they speak not so much to display frustration as to solve problems; and
4. Alter behaviors and carry on adaptation in a rational and orderly way.[2]

It is interesting to note the similarity of the personality factors measured in the MPQ; the five categories of emotional intelligence, as discussed by Raye Kass in her contribution to this volume (see "A World Waiting to Be Born," p. 53); and some of these characteristics associated with effective conflict management.

Team Decision-Making Processes

The Mars One team will have a lot of discussions about what to do and how to get it done. There are several key principles to help these discussions be productive that deal with *task* and *process* in group dynamics, the *creative use of conflict*, and *delegation of responsibility* between decision-making sessions. These principles will help the group handle conflict and function in a way that manages the emotions, negative and positive, that are likely to emerge.

Task

A group's task is the stated purpose or outcome of its interaction: the *what* of its activities; its results. It helps to have the following:

1. *Clear objectives*—team members know why they are meeting and what they hope to achieve.
2. *Focus on outcomes*—team members demonstrate commitment to the group's outputs.
3. *Achievement of results*—the group achieves its objectives in a timely and effective manner.
4. *Sharing of the load*—the workload is distributed appropriately among the group's members.
5. *Learning from experience*—the group systematically evaluates its results and applies the lessons learned to improve its output.

Process

A group's process is the way it goes about its business—the *how* of its workings. An effective group needs the following:

"If I'm on a team and we have a common purpose, I'd do my best to help achieve our goal, and I will even sacrifice my own convenience to make this mission a success."
—Ali Rasoulzadeh Darabad, Mars One applicant

1. *Common purpose and principles*—operating principles and a purpose that is shared and overt
2. *Role clarity*—roles that are clearly defined and understood by all members
3. *Climate*—members who feel it is safe to be honest; who trust and respect one another and hold one another in

high esteem; and are unified and supportive of the group's decisions

4. *Consultation*—members who express their opinions freely and fully, and with moderation
5. *Listening*—members who feel they are listened to and that their comments have an influence on the group's progress
6. *Self-discipline*—members who avoid giving or taking offense
7. *Detachment*—expressed opinions that become the property of the group, and when a differing opinion is expressed, it is seen as another contribution to a shared exploration of the issue
8. *Conflict*—to view a conflict as a clash of opinions that serves to shed light on the issue being discussed rather than as a clash between the people who voiced them
9. *Order*—shared responsibility for ensuring participation; when the group at large fails to maintain order, a designated member of the group uses his or her responsibility to keep things on track, and the group's decisions are clearly defined and recorded for action

Creative Conflict

The constructive and creative use of conflict is essential for group survival and growth, particularly in diverse and turbulent low-predictability contexts. In his book *The Conflict-Positive Organization,* Dean Tjosvold says an effective team has the following characteristics:

- Members *value their diversity* and recognize the creative potential in constructive conflict. They look for opportunities to voice their different views, discuss frustrations, and work to make their relationships productive.

- Members seek *mutual benefit.* They understand that they have mutual interests and seek common ground. They are all committed to pursuing a shared vision and creating an environment that is fair and facilitating for all.
- Members feel *empowered.* They are confident they have the mandate, opportunities, and skills to manage conflict.
- Members regularly *take stock* and *reflect* on their handling of conflict. They realize that becoming conflict-positive requires continuous experimenting, feedback, and group-directed improvement.

Delegation of Responsibility

The effectiveness of a group depends on how well it delegates responsibilities and supports its members' activities from one decision-making meeting to the next. Members should feel confident and free to carry out their responsibilities within a clearly defined framework, such as follows:

1. *Clear guidelines*—they know when they are free to act and when they should seek guidance from others.
2. *Clear task definition*—they know what is expected of them between meetings.
3. *Available assistance*—they have ready access to the help they need between meetings.
4. *Freedom, safety, and support*—individual initiative is encouraged between meetings and the group supports its members' efforts to carry out their responsibilities.

If the members of the Mars One team systematically study their interaction using these analytical tools, it will foster their unity, harmony, creativity, and, ultimately, the success of the mission.

This article has presented a checklist of elements associated with improving the cultural "soft skills" at the root of human relations and effective teams, many of which may be relatively well known in some circles. The article highlighted that we learn to see the world using lenses acquired through our socialization process, and this pattern is a presumed reality for each of us. Likewise, the rules we learned in childhood that propel us from the depths are almost completely hidden to us but can occasionally be brought to the surface by checking the sources of the emotions that influence our actions.

A strategy to improve human relations is to become emotionally self-aware and catch our emotions before they become potentially problematic behaviors—to become progressively more emotionally intelligent. If we trace these emotions back to their source in the deep levels at the base of the iceberg of our personal culture, we'll find the hidden rules at their root. We can then decide whether we want to keep those rules or change them to better suit the world in which we want to live. We can decide to keep the blind imitation of the patterns of thought we inherited from the societies in which we grew up, or we can consciously decide to change those parts of the foundation of our being so we can better contribute to a more harmonious and diverse world.

> "I think about the existence of life a lot, and if there is any purpose of it at all. I believe there is no good answer to that question but we create the purpose ourselves. But I do believe there is an explanation to the question—we just need to figure it out. To do so I believe we need to look for answers in space, to search for other life forms of any kind."
>
> —GUSTAV ERIKSSON, MARS ONE APPLICANT

However, not everyone is familiar with this type of process. When in Cairo some years ago, working on the start-up of a new hospital, I used a team analysis instrument based on the categories for analyzing group process listed earlier to help a group of doctors become a more effective team. They told me during the session that it was the first time they realized it was possible to analyze a meeting, and they did not know it was reasonable to expect team members to control their emotions and manage their behaviors so the group could be more united and productive.

They seemed to assume that existing individual characteristics were natural and fixed expressions of personality and not amenable to being analyzed or controlled—that they were a part of human nature that was hardwired and not able to be changed or improved. I question this assumption, and point to a critique of the pattern of human relations depicted in movies like the *Star Wars* series. If humanity were to maintain those characteristics while their technologies advanced, as shown in those films, the human race would likely self-destruct well before reaching that level of technological advancement. A fundamental change in the character of humankind would be required to develop and properly manage that technology. The Mars One mission is an opportunity for humanity to learn how to advance on the technological level as well as on the deeper dimensions of our human-relations behaviors, values, and beliefs—I wish it well.

Andy Tamas *has been offering intercultural training and organization development services since 1975, and more recently became engaged in strengthening governance in fragile war-torn states such as Afghanistan, Iraq, and Yemen. His earlier work with intercultural relations and fostering harmony and diversity in rapidly changing societies contributes to the foundation of his nation-building activities.* *He would like to thank Michael Bond for his constructive comments on an earlier version of this article.*

What role does age play for the settlers on Mars?

AGE AND AGING ON MARS

MARS 100

"Age is nothing more than a number pointing to generational origin. Everyone will age at the same rate. They will grow together. They simply have different starting points."

"It came as a surprise that fifty is different from forty. So I expect my midsixties will be different, too."

"At thirty-eight percent of Earth gravity, Mars will be great for our joints!"

"Growing old on Mars is the best thing I can imagine. A constant life of knowing that you are doing something fantastic. This is what I want to do with my life."

"Age is a number that comes with a lot of attachments—differences in thinking, fitness, deliberations, and so on."

These were a few comments made by some of the final one hundred candidates for the first settlement on Mars. The team that goes to Mars will be diverse in gender, cultural background, and age. This is intentional. The organizers of the Mars One mission believe that this diversity will help the team bring greater resources to solving problems. The diversity also reflects humanity, and perhaps will result in a more useful experiment for learning about the needs of future colonies that may develop on other planets, moons, and asteroids.

For this section, we set out to discover what the people who are ready to leave Earth think about age and the idea of growing older on Mars. What benefits do they see in an intergenerational settlement experiment, and what problems do they expect? What aspects of aging on Earth do they think they will miss? What kind of culture do they want to create, relative to age?

We collected thoughts from thirteen Mars One candidates who ranged in age from twenty-one to sixty-one and represented a mix of cultural backgrounds.[1] In this section, we hope to show you different perspectives as well as points of concurrence. The topic of aging on Mars is largely one of conjecture. To be sure, we know something about the aging process on Earth, but just a bit about the effects of zero gravity on the human body. Everything else is extrapolation and guesswork. We will focus instead on the views of these potential settlers because, ultimately, they will shape the first society on Mars.

With the mission departing in a little more than a decade, the youngest of the potential candidates could be in his or her early thirties, and the eldest past seventy—close to one-and-a-half generations apart. Though family, country, culture of origin, and personal experience play an enormous role in shaping an individual's perceptions of the world, it's helpful to realize that the youngest potential settler generally came of age at a time when computers, mobile phones, the internet, and rapid technology deployment were standard. National and other boundaries were blurring, and smaller wars and skirmishes made enormous differences in local life even as the world moved toward a more global sense of itself. For the eldest of these candidates, life was quite different: The Cold War created ubiquitous global tension even as it spurred the space race and other technological developments. Communication was slow by comparison (hence the term *snail mail*). The world was less connected. The idea of "multitasking" barely existed.

On any number of levels, the settlers will arrive on Mars with different worldviews. Yet they will have trained together for the decade needed to learn survival skills for inhabiting a hostile planet. This gives them ample time to come to grips with their dissimilar backgrounds, mine the benefits they find in those variations, and figure out where generational and other differences might threaten the team.

Strengths at Any Age

What do the potential settlers see as their strengths relative to a given age? Regardless of a candidate's particular age, he or she was more likely to describe positive attributes for various age groups and less likely to dwell on problems; a younger candidate might mention many attributes of the elder candidates, and elders ascribed many benefits to the younger settlers. Some of the candidates had already had the opportunity to meet one another in various contexts, and one noted, "Age is blurred in this group. Everyone has youthful enthusiasm."

All candidates recognized many strengths to be anticipated from different age groups. Most felt that these would change over time, although a few candidates did not think age would result in many changes.

Thirties to Early Middle Age

Some younger applicants felt that younger people have had less time to be molded by their culture and society and therefore are likely to be more flexible in their thinking.

One noted that younger team members would help to activate widespread social support for the mission: "My generation hasn't seen any huge progress in human space exploration. We've had some remarkable unmanned operations, but we are not on the human track. So this means we are really ready to ignite the effort to do something new. A good idea that is well managed and maintained has the potential to go far with my generation—the generation that is just finding its place in society. We are more likely to accept the idea that if something has merit, it can work. We don't need an institution to support something."

Some noted a greater willingness among younger people to take risks. Related was the idea that young people tend to be more idealistic and this idealism made them willing to sacrifice for a cause. One younger applicant said, "Survival does not matter. We want to make changes."

Younger people were viewed as able to learn faster and as possessing more ambition. Said one younger candidate, "This is an age when you want to leave a legacy: You are ambitious, full of energy, don't want to give up, and have stubbornness, but in a good sense."

Candidates anticipated greater solidity in all aspects of life upon entering early middle age. One said, "At my current age, as compared to when I was twenty-two, I've had a lot of time to reflect and mature. So by the time we leave, I'll really understand my own psychology, and this will be beneficial to the team. I'll spend half my life on Earth and half on Mars."

A younger candidate noted, "I think on Mars or on Earth the process of aging will be the same. As you age, you will gain more wisdom and be more patient, because you already have experienced a lot in this life and are not so emotional about things. You will be more smart in terms of the psychology of people. You will expect less from people than when you are young and idealistic and expect people to be the best all the time. So, by age forty or fifty, you are more relaxed about some things."

Middle Age and Older

The strengths expected for this age category were a greater bank of experience, a more thoughtful approach to problems and crises, and a more empathic approach to interpersonal challenges. One candidate said, "As we get older, we come to understand people better. We understand more of other people's viewpoints."

Older candidates felt they might be able to offer more perspective on and patience with the challenges that would face the team. One candidate who had experienced numerous crises and life-threatening situations, including war, noted, "I take things calmly and I don't panic."

Some candidates—young and old—felt that because older candidates had grown up in a time before widespread technology and disposable consumer goods, their age would be an asset when problem-solving was needed by the settlers. One noted, "Sometimes there are simple and practical solutions to problems that we rely heavily on technology for today. This is a generalization, but people who grew up before the advent of today's technological advances (the internet, mobile phones, 3-D printing) bring a bank of experiences, knowledge, and understanding of different systems that would be beneficial."

Several middle-aged and older candidates noted that they would be especially valuable for extra-habitat excursions because they would have a shorter probable life span on Mars and could therefore tolerate more radiation. Because it would take sixty years living on Mars to reach the European Space Agency's radiation limits, and older settlers would likely only live fifteen to thirty years after arrival, they would not "max out" their exposure as younger settlers might.

Several candidates felt that age would be an advantage when welcoming new teams of settlers. "Those of us who have been on Mars the longest will have a lot to offer in terms of what we've learned. So we can grant this to the newcomers, which will be an important role for our aging members," stated one thirty-something candidate.

One candidate braved the topic of sexual relations among settlers, noting that "one of the benefits of being older is it is easier to push sexuality to the background than when you are thirty. Someone in their sixties is going to have a better understanding of the complications that can arise with an intimate relationship. Repressing sex doesn't work. We can't afford a lot of the normal dynamics in romantic, familial, and sexual relationships. We can't have the benefit or the hindrance of secrecy. An elder can say, 'Let's get things out in the open. And let's be cautious because, given the constrained space and limited company, if feelings change and the relationship breaks we will still be working together for the foreseeable future.'"

Regardless of age, candidates looked forward to contributing to their final days. For example, one candidate said, "On Earth, you can retire. On Mars, you will have to work until your dying days—to always be strong and stay young. We will all lose some energy over the years, but we will be forced to give our best all the time."

Age-Related Challenges

Candidates also discussed some of the weaknesses related to any given age. Some of these are simply the obverse of those noted above: Younger people may be less cautious and may inadvertently put themselves or others at risk, while older ones may not think and solve problems as quickly as younger ones, which is risky when rapid solutions are essential. With a few exceptions, most expected to see some deterioration in reflexes, strength, and speed of thought, along with other physical changes associated with aging. One noted that the controlled diet, regular exercise, lack of diseases, greater medical attention due to frequent medical checkups, and lower gravity may compensate for the wear and tear imposed by the harsh environment. One of the older candidates said, "I am not sure how the different living conditions will impact aging. You can adapt gradually. Nothing will change overnight. It's bending versus breaking. If you bend gradually, you'll do well."

Two candidates noted that they felt just as physically and intellectually fit in their fifties as they had earlier in life, and fully expected to maintain health with age based on the way they'd observed other people from their cultures of origin aging. One said, "Some individuals or cultures see aging as a negative process. So I think, before we form a team, we need to talk and establish a mutual understanding about what aging is. And if I can show that aging is not to be feared but something to embrace and enjoy, it will affect young members so that as they get older they will have more positive views toward aging."

In contrast, one younger applicant was concerned that the closed environment might make maintaining good physical fitness more difficult as settlers aged: "It may be difficult to stay motivated to stay in shape on Mars—motivation may actually diminish."

Another concern is that the ability to work in a spacesuit will be lost should physical strength diminish. "All the current suits require a lot of physical effort and strength to work in," said one candidate. "So as I age, I won't be as strong. The flexibility of the suit is a very big concern for me."

A number of potential settlers interviewed were excited about the prospect of aging on Mars, particularly as a personal experiment. For example, one said, "I want to see how I change. Am I going to change or not? This whole thing is an experiment, and my life is an experiment, too. Whatever happens in my life, I like to enjoy it."

Another candidate said, "It's going to give us the most unique perspective of aging in all of human history. Here on Earth, you have kids and grandkids, and you focus on them. On Mars, you are focusing on giving the whole species a new frontier. And that is a very different way of aging. You want to do things right. You want to instruct new generations on how to live and be happy in a hostile environment."

Another candidate stated that the closed community will make the loss of a settler even more emotionally painful than a loss on Earth: "Emotionally, as time passes on Mars, you will become increasingly attached to the other settlers. This may make letting go due to a tragic event much more painful. You will need to be attached more than you would on Earth, but be emotionally prepared to release those connections so that the mission is not compromised."

Reflecting that there could be as much as forty years between the ages of the eldest and youngest first settlers, a younger candidate remarked, "It may be difficult to see some partners grow older while I'm still in my prime."

A handful of candidates from all age groups brought up end-of-life issues. Chief among their concerns was the fear of becoming a burden to other settlers, and several brought up the idea of self-directed euthanasia. One said, "When a team member gets frail, well, this is something we'll be trained on and discuss over ten years of preparation. Before we go to Mars, we must understand and sign do-not-resuscitate orders."

Another was more direct: "I believe that at some point every person has to assess whether they are contributing to or becoming a drag on the mission. But I would not want my teammates to have to make the decision or to feel some sort of survivor's guilt. The medical staff or the psychology staff on Earth would be empowered to say, for the good of the mission, it is time to give X the black pill."

One candidate foresaw a change in attitude toward end-of-life: "I think in the early days people will have traditions brought with them from Earth, whether they are family related, religious, cultural, et cetera. I would assume that if the settlement is capable, each individual will get their wish for how the end of their life is handled. I think that once the settlement has grown considerably, we may need to decide on a way to deal with end-of-life issues that is efficient for the settlement and agreeable to its inhabitants, but that is a long way off."

Celebrations and Losses with Age

The candidates interviewed were asked to think of themselves at the age of eighty (in Earth years), having lived on Mars from one decade to nearly five decades. What would they celebrate, and what would they miss?

Several said that they felt the toughest adjustments would come during their first decade on Mars. Those who would live a long life on Mars thought that, after the adjustment period, living on Mars would feel normal. "At eighty, I will have lived forty years on Mars. If you are going to miss things, it will be in the first years," said one candidate. Another said, "The first ten years you'll spend missing things on Earth, but after that, it'll be home. At eighty, you'll reflect on what you've done on Mars."

Most candidates felt that they would miss being with their circle of friends and family and would miss being with these loved ones when they passed away. One said, "My parents are not getting younger. It will be tough not to give a last hug in life. But this is something we must get used to." Another candidate said, "I'll miss having kids and grandkids. At the same time, we are exploring a whole world, which is fascinating. We won't have traditional things on Mars—seeing kids go to prom, getting married, all those kinds of things. It will be hard, but the human brain has protections. When you grow, you remember more good than bad. This is natural: to protect ourselves so we are not overstressed. This will also occur on Mars. We will all be homesick in our own way. But then this protection will kick in. You will remember your Earth life, but it will be more like a dream. Reality is *here*, on Mars. Worrying all the time that you can't see or touch your family will cause pain, so you will not do it."

One said, "I will be missing the history I might have had, which might include children, a spouse, house, dog, et cetera." Another said, "There is comfort to being able to sit around with your peers and reminisce about 'when I was a kid.' So the opportunity to do that on Mars is limited. I would miss that. And I would miss having a dog."

One potential settler felt that the need for family would be replaced by the relationships formed with incoming settlers: "I think that by the time we've actually departed, whichever teams we are in will be a family. We will be close to the other teams that will be coming, too. So we will have a lot of time to appreciate each other. A lot of us are getting together already. This is the nature of humanity; we find love and connection wherever we are."

Several candidates discussed the differences in expectations and treatment of older adults in Eastern and Western cultures. One bicultural candidate suggested that aging on Mars might be preferable to aging on Earth in Western society: "I don't think we treat our elderly that well. I feel sad that as you grow old you disappear and become more invisible. This is different in Asian cultures, where we have roles as matriarch or patriarch. But in Western society, we see this worth diminishing as you age. This won't happen on Mars because there are not enough of us. And we are interested in learning about aging on Mars, what happens with low gravity, our diet, radiation, et cetera, so we will be useful specimens throughout all stages of life."

Many noted that reaching eighty on Mars would be a great cause for celebration. "Aging on Mars will be a completely different scenario," noted one candidate. "The people that live their last days on Mars and are dying of old age—eighty, ninety years—will have a feeling of success, and can say, 'I am passing on now, but I have expanded the limits of our species.' Here on Earth, an average elder dying may look at his life and ask, 'Have I done right, wrong, reached my goals?' Seeing yourself on a different planet should mean you die happy, leave a legacy, inspire people back on Earth, maybe even inspire peace, because those on Earth have seen people getting along so well on Mars."

Said another, "Oh my goodness, I'd be a happy old lady. I'd want to scream and shout, 'Guys, anything is possible. Don't ever be limited by what you see other people doing.' That's a drive in my life, and a drive for why I want to go. I'd be well lived and well used."

Intergenerational Teamwork

Research on intergenerational teamwork is mixed, and typically seems more focused on intergenerational work within the same culture. Much of it is related to workforce management in Western corporations. Because the Mars One organization has purposely sought candidates from around the globe, potentially vast differences among team members are to be expected. These diverse teams must surface and utilize the strengths of each individual.

Within this context, the candidates interviewed were asked to talk about some of the strengths and weaknesses of working across generations.

Most were confident that the ten-year training period would help them overcome problems related to generational differences. They expected some issues, but also appreciated the benefit of mixed-age groups in both work and social experiences. For example, one candidate said, "With too much difference in age, it can be hard to get along. But at work, when there is a mix of ages, like sixty, fifty, forty, thirty, then that helps the group blend as a whole. The various ages help connect each other. We need younger and older people. We need younger people to do the physical work, even in their forties and fifties. But the older, deliberative person will also be important."

An older candidate stated, "You are going to be training intensely over a decade with a group of people. You'll be exposed frequently to the cultural norms of those generations. And I'm hoping that everyone is open to absorbing what is good and enjoyable about that. So I don't think the thirty-year-olds will think of older settlers as old fogeys, nor will I think of the young settlers as disrespectful urchins, since we'll work together every day. We'll watch movies together, share books, and mix more actively than most people do normally."

Conflicts were expected by a few, not due to team goals or ages but rather lifestyle. One candidate said, "Habits require more adaptation. If someone is not organized, someone is very organized, someone has good hygiene, another is sloppy—those are the things that will require more adaptation among team members. Once we are beyond that, it is all about training on the same path."

A younger candidate shared a similar perspective: "As far as work goes, we all have a common desire to make this a success. So as far as working toward the mission, I don't foresee too many conflicts. But issues may come in downtime, when you are trying to connect, grow, become a unit. People of similar age have a similar background, so you can make culture a reference that helps you grow bonds."

All candidates easily named many benefits to living and working with a mixed-age group:

- "You can feed your knowledge from all sides with people of different ages, cultures, parts of the world. Older generations can show the point of view of a completely different world than the young ones grew up in. With a long, one-way trip, you need variety."
- "It is just like a family structure on Earth, in which there are different ages. In this way we can set up a family

structure and survive well. Age, maturity, and experience all are necessary to construct a social order. In this regard we will form a new order of society."

- "People of similar age have only seen similar solutions. You are exposed to more problems as you age. That's why, in universities, professors are so much more experienced than students. Young ones have a problem, they can talk to an older professor who has faced a similar type of problem before. So problem-solving becomes more effective in a team with mixed ages."
- "A younger team member may have a brave idea of doing something new and be ambitious and emotional about it. He might not see side effects. But an older, more experienced crewmate with more emotional intelligence might balance this and help the team consider other sides of the idea. The wisdom that comes with age is a strength. Different perceptions come with different ages and these can bring pluses and minuses to the team."
- "The strengths of a mixed-age group will be having a function of 'gas pedal, brake, and gear' in the same team. The elder generation probably can listen to the younger ones patiently and provide different perspectives. The problem will be different energy levels among the generations. This could lead to frustration if we don't communicate well."

One candidate summed up the unique aspects of a mixed-age settlement team quite well:

> I feel that multigenerational teams are more creative problem-solvers, and share the richness of their diverse experiences and perspectives in such a way that can improve one's appreciation of life in general. There will likely be challenges in communication and deep comprehension of

language between generations—both on Earth and Mars. The community on Mars will be small. People will be completely reliant on each other for survival, so it will be important for the crews to overcome these challenges. The same challenges will be present due to cultural and gender diversity, and the same richness and rewards will also ensue.

Creating a New Culture

Ultimately, the people who engage in this experiment will be on their own, at least for some time. They will create their own culture. What will this culture be like? How will age and generational identity play a role?

Many candidates said they desire a culture that features qualities such as respect, tolerance, gentleness, nonviolence, scientific inquiry, and accommodation, as well as the importance of respecting a person regardless of age. Several brought up the idea that should the settlement have children they should be educated free of bias, expected to contribute to society, and taught to respect the contributions of other people regardless of the contributor's age. A number wished to see a settlement that based decisions on rational discussion of merit rather than tradition or faith. In various ways, they expressed the idea that because Mars is a hostile environment, settlers will need to think of others first, and that decisions should be made in terms of how they will impact the whole rather than the individual. Some described this as a particular antidote to the self-centeredness of the industrialized world, and one hoped that this new culture could serve as a peaceful example for Earth.

Several stated that as people age, they were due respect for their experience and greater knowledge, but that respect did not transfer automatically to authority. They expected that due to the fragility of survival for the foreseeable future, Martians will not cease working as they age, and the culture of the settlement will expect everyone to contribute throughout life. For example, one person said, "I believe any generation of people should be treated with respect and held accountable for their actions, but also be given a voice to contribute to the community. Anyone of any age can have good ideas and contributions, but as members of a smaller community in a precarious environment, people will have larger and more vital responsibilities."

One candidate added a caveat about settlers' attempts to design a culture: "I really do want to make sure that the society we build is free of any kind of bias. I want people accepted as individuals on their own merits, including their age at any given time. That means we give respect, but not deference. I don't want somebody saying, 'You can't do that because of your age or your gender.' That is not the society I want to create. However, American history is littered with all sorts of intentional communities that formed and then died out or dissipated. While it is my hope that we will create Martian society with deliberation, that we'll look at many philosophies and economic systems, it is hard to say that, in the long term, it will be a successful endeavor. But I still want to try because that is the way I am."

What role does gender play for the settlers on Mars?

MEN ARE FROM MARS, WOMEN ARE FROM MARS

Gender Dynamics in the Mars One Project

RONIT KARK

In his bestselling 1992 book, *Men Are from Mars, Women Are from Venus*, John Gray writes that men and women are so fundamentally different, psychologically, that they might as well be from different planets, and that most of the common relationship problems between men and women are a result of these psychological differences. Men, for example, provide information, while women express emotions. According to Gray, men respond to stressful situations by offering practical solutions to problems; women are not necessarily interested in getting practical solutions to problems, but want mainly to talk openly about them. The book even suggests that each gender has a distinct language and offers a Martian/Venusian Phrase Dictionary to translate commonly misunderstood expressions.

Of course, many researchers have challenged Gray's notions, asserting that men and women are more similar than they are different—and Gray himself has updated his work on gender along similar lines. Or as the organizational psychologist and Wharton Business School professor Adam Grant said in the *New York Times* in 2014, "Instead of claiming that men are from Mars and women are from Venus, [it's] high time to recognize that we're all from Earth."

That said, no one (on any planet) denies that men and women confront different social realties and challenges. So what happens when men and women from Earth *go to* Mars? Mars One is attuned to the advantages of gender diversity and has committed to sending gender-balanced teams to Mars. However, this commitment may pose challenges at different stages of the project, from astronaut selection to the interaction of team members themselves. Still, with careful and insightful attention to gendered issues and a heightened awareness of the challenges involved, men and women can not only work together to successfully travel to Mars, but also build a community there in which both men and women contribute to a meaningful life, in a society where they have equal opportunity to exercise power and influence.

May the Best (Wo)Man Win: Candidate Selection

In a 2014 article in *Slate* called "Why the First Mission to Mars Should Only Have Female Astronauts," Kate Greene argued that women should be the first to Mars because of basic economics. Women, she contended, are cheaper to fly than men because they expend fewer than half the calories of men and weigh less, reducing food and fuel costs. Up until now, however, the number of women who have traveled to space has been limited.

In 1963, aboard the *Vostok 6* spacecraft, Soviet cosmonaut Valentina Tereshkova became the first woman to travel into space, just two years after the first man. However, it took more than twenty years before the first American woman traveled to space—in 1983, when astronaut and physicist Sally Ride served as a mission specialist on the space shuttle *Challenger.*

NASA reports that since 1961 a total of 534 individuals have flown to space: 477 men and 57 women. These women and men had somewhat different qualifications. For example, the women had higher levels of education, possibly suggesting that they needed more qualifications to be chosen over male candidates. Specifically, to date, female NASA space station astronauts have had almost twice as many doctorate-level degrees as their male counterparts (50 percent versus 28 percent). Male NASA space station astronauts have had more military experience (73 percent versus 39 percent).[1] The women chosen for NASA's space station tours have been an average of two years younger than male astronauts. And although there have been no significant differences in the percentage of male (76 percent) and female (69 percent) space station astronauts who were married, a significantly greater percentage of male astronauts had at least one child (67 percent versus 38 percent); and, overall, men had more children than women, possibly indicating that it may still be more difficult in our society to be

"A human existence on Mars represents so much possibility, all the best of humankind boiled down to one endeavor. Mars is a dream that can unite people across petty borders and cultural differences. Mars can show us better ways to live to the true potential of our magnanimous nature. Mars is necessarily the very peak of our scientific and intellectual endeavors to date, and opens the doorways to even greater exploration of the universe.

"So for me, Mars is both the practical solution to so many of Earth's problems, and at the same time it is the noblest aim for an almost inexpressible human reach."

—Mars One applicant

a "mom in space" than a "dad in space."[2] A new class of NASA trainees, announced in 2013, is the first group within many years to include equal numbers of men and women. Four men and four women were selected from a pool of sixty-three-hundred applicants that included the highest percentage of female candidates ever.[3]

Mars One's stated intention is to select a gender-balanced group of four astronauts. This may be a challenging process for several reasons. First, given the historical precedents, most people tend to think of astronauts as male, not female, and thus may be less likely to see women as fit for the role. Second, stereotypes around gender and power significantly influence the ways in which men and women are evaluated by organizations, particularly for roles, like space travel, that involve risk.[4]

Identical CVs for the same jobs are reviewed and evaluated differently when the name of the applicant is a woman's versus a man's. Once hired, women are held to higher standards of performance than their male counterparts as a result of a perceived incongruence between their gender roles and leadership roles.[5] This phenomenon is particularly evident under three conditions: in roles and organizational environments that are defined as stereotypically masculine (e.g., firefighters, police, banking, etc.); in workplaces with large numbers of men in comparison to women; and in situations in which most of the assessors are male.[6] As can be seen from the NASA statistics above, the role of an astronaut is traditionally a highly masculine role and the work environment is clearly male dominated.

"It is my belief that we as intelligent beings have a duty to ensure our own survival and the survival of life in general. As long as we are confined to a single planet, we are vulnerable to extinction by a single event."

—Mars One applicant

The gender selection bias of astronauts in organizations such as NASA goes back to the beginning of the space age. Instead of processing thousands of applications from the public at large, NASA decided it would be easier to sort out the military records of a few hundred active-duty military test pilots. In 1958, President Eisenhower approved this pragmatic plan. So a select group of military test pilots were called in, briefed on Project Mercury, and offered the chance to apply to be astronauts. Applying to be an astronaut was by invitation only. At that time, there were no women military test pilots because the US military didn't accept women into pilot training. So the question of whether to accept women as astronauts never even came up. As later groups of astronauts were recruited, the criteria were loosened, but the choice of military test pilots as the first astronauts had set the pattern. Not until the space shuttle began flying did NASA conceive that not all astronauts had to be jet pilots. And at that point, sure enough, women started becoming astronauts. Nonetheless, women did not start showing up as shuttle pilots—and hence, eventually, as shuttle commanders—until the US military began training female pilots.[7] And our concept of *astronaut* remains primarily male to this day.

The tendency to assign men and women differently to roles has to do with perceived incongruities between how women are perceived and the ways in which male sex-typed roles—like astronauts or management—are perceived, as captured in the known phrase "think manager—think male." Such incongruities manifest in lower performance evaluations for women and in more difficulty for women to advance to desirable positions in roles that are perceived as masculine. In other words, we would expect female candidates to receive lower assessments as astronauts because their gender roles (as women, mothers, wives) are incongruous with the social role they would fulfill as astronauts.[8]

This leads to a dilemma women face in such masculine typed roles: The prescriptions for the female gender role stipulate that women should be especially *communal*, and the prescriptions for an astronaut role stipulate that they should be especially *agentic* (i.e., assertive and proactive).

The resulting dilemma is that highly communal female candidates (warm, considerate, sensitive) may be criticized for not being agentic enough and not properly taking charge, whereas highly agentic female candidates (dominant, forceful, decisive) may be criticized for not being feminine, communal, or nice enough. And violating either the "proper astronaut" or "proper woman" stereotype, research indicates, can lower evaluations of women and their performance, leading women to experience the double bind.[9]

This links to Ambivalent Sexism,[10] which describes stereotypes and prejudices against women as a combination of two opposing but complementary voices that contribute, through either benevolent or hostile discourse, to the conservation of the gender status quo. The prejudices at play here are twofold: "benevolent sexism," a paternalistic ideology that views women as dominated by men and primarily suited to low-status, traditional roles; and "hostile sexism," a paternalistic ideology that views women as a threat, sparking fear that their supposed attempts to reach higher statuses are at men's expense. Hostile sexism is directed primarily at strong, opinionated "modern" women.[11]

In a study of an assessment center for managerial positions in a masculine security organization in Israel, performed in 2014 by Varda Wiesel and myself, women were found to experience benevolent sexism and to be positively evaluated when they were perceived as weak, and to experience hostile sexism and be more negatively evaluated when they were perceived as powerful. One of the female evaluators in the Assessment/Selection Center explained the generous evaluations given to female candidates this way:

> It's harder for a woman to succeed in this system, it's harder to succeed as a female manager . . . At the Assessment Center you can't ignore it. When one woman is sitting with five men . . . especially when a woman is alone . . . these are places where we're [the evaluators] lenient . . . [I]t's a factor, "let's help her out a little so that she'll be equal among them."

In contrast to this generous evaluation, assessors stated that they took a hard line with female candidates who adopted management styles that they considered "masculine." In other words, women who were perceived as weak and conforming to expected stereotypic feminine behaviors were rewarded, whereas women who behaved in a non-stereotypic gendered manner experienced hostile sexism and received lower ratings. As one of the assessors said in an interview:

> There's something unappealing to the eye and the heart in seeing a woman attempt to manage in a masculine style, I think it confuses us, it doesn't seem natural to us. I think it influences all of us . . . We don't really like to see a woman behaving in a masculine way.

Gender-conforming female candidates receive higher evaluations than female candidates who exhibit "masculine" management characteristics. They are also more likely to be promoted to management positions. Women who conform are more likely to accept the rules of the hierarchical/hegemonic system and are less likely to challenge the existing gender order, which preserves and reproduces the male power structures.

Thus, although many organizations and assessment centers use supposedly objective sorting tools and criteria that attempt to minimize hierarchical gender divisions and stereotypical gender thinking, they can still end up supporting a regendering process—a process that preserves and re-creates gender distinctions that place women at a lower status than men. It's an issue Mars One will need to take care to avoid.

It is interesting to look at the current list of qualifications for astronauts published by Mars One. The qualifications appear gender neutral at first glance. For example, the organization is looking for characteristics such as "resiliency," as shown via a "Can do!" attitude, and "persever[ing] and remain[ing] productive." They also look for "creativity/resourcefulness," defined as "You have a good sense of play and spirit of playfulness," and "Your humor is a creative resource, used appropriately as an emerging contextual response."[12] However, underlying these seemingly gender-neutral attributes may be a gendered substructure. Because originally astronauts were almost exclusively men, the role of an astronaut was designed with men in mind. As Joan Acker notes in "Hierarchies, Jobs, Bodies: A Theory of Gendered Organizations," her famous work on role selection and design, "Gender is difficult to see when only the masculine is present."

The notion of a "resiliency" and "Can do," as well as a "sense of play" in the Mars One qualification list, for example, may be more prevalent among men. In her recent book *Overwhelmed: Work, Love and Play When No One Has the Time*, Brigid Schulte describes how researchers show that women lack leisure time; instead, they fill their time with work and chores, and do not make the time to play. "The fright of human history," she notes, "can be boiled down into three powerful words: 'Women. Don't. Play.'" Thus, the seemingly innocent requirement for the Mars One crew to be playful and have a spirit of play may, in fact, not be gender neutral at all—it may give men an unintended advantage. If special caution is not taken in the selection process, through the requirements criteria and assessment process, there is a chance that one gender group (more likely men) will gain an advantage, despite Mars One's best intentions. This is one of the things that Mars One evaluators must keep in mind while selecting women for their project, if they are looking to start a society free from these prejudices and gender expectations.

From Glass Ceilings and Glass Walls All the Way up to Space: Leadership and Gender

Although women in different nations have achieved more access to leadership and influential roles now than at any other period in history, equal representation of women and equal access to power, specific roles, visibility, and prestige remains a distant goal. Given that sex discrimination is illegal in many industrialized nations, its continuing presence may seem puzzling. In many contexts, blatant expressions of discrimination have disappeared. Yet, as outlined in the previous section, discrimination continues in subtle and often unconscious forms. People who behave in discriminatory ways are likely to believe that they are merely choosing "the best person for the job" or otherwise acting in an unbiased manner.

Glass has been used as a metaphor for many examples of female exclusion, including the "glass ceiling," "glass wall," "glass escalator," and recently the "glass cliff." The term *glass ceiling* first emerged in the

1980s to describe the experience of women who face
difficult-to-recognize barriers that prevented them from
highest ranks in organizations, and unfortunately it remai
today. Since many of these barriers are invisible, women (and
to perceive the barriers that prevent their advancement, and this makes it even more difficult to challenge, overcome, or eliminate them.

In addition to and in line with the gendered stereotypes and expectations we have already explored with reference to communal and agentic roles, one of the most obvious answers to the questions of why women hit a glass ceiling and why more men than women occupy leadership positions is that, for most women, the path to workplace advancement involves negotiating trade-offs between work life and domestic responsibilities. As noted previously, this is the likely reason women astronauts had fewer children than men in the same positions. In June 2013 the CBS show *This Morning* conducted an interview with astronaut Karen Nyberg, who at the time was on a six-month mission in the International Space Station. The title of the segment was "Mom in Space," and it started with the following question: "You left a three-year-old son, Jack, at home. You missed his first day at preschool. How is that? Being away from your young son?" Those are important questions. However, we rarely if ever ask them of men. And this certainly would never be the main focus of an interview with a male astronaut. As long as people have the stereotypical expectations and perception that women are the ones responsible for children, women will be limited in their opportunities to achieve and hold power, and men will be limited in their opportunities to be fully active dads.

The concept of the glass wall refers to an invisible barrier preventing women from moving laterally within an organization, many times to core areas. This forms occupational segregation that restricts women's access to certain types of jobs (or agencies), or that traps them within certain types of jobs (or agencies). For example, 10 percent of NASA's astronauts are women. However, 20 percent of the astronauts who have flown to the International Space Station have been women—double the percentage. These numbers may suggest that space stations, s a domain, are perceived as somewhat more fit for women than other areas of space travel—such as the more often emphasized and lauded space exploration.

The glass metaphor coined for men's experience—the glass escalator—describes the way men are more likely than women to be accelerated into influential leadership and management positions, even in fields that are associated with female roles. Whereas girls and women, as subordinates in the gender order, are allowed and even encouraged to strive for what used to be solely the province of men, boys and men are not similarly encouraged, nor even allowed, to move into the province of the feminine. In fact, when men work in traditionally female fields, we see a tendency for men's superordinate status to be restored via fast promotions.[13]

During Mars One's selection process (and in the eventual community founded on Mars), teams will need to guard against most if not all of these issues, lest they limit female candidates' ability to hold influential leadership roles based on their character and abilities rather than their gender. The glass ceiling means the female members of these teams may struggle to gain leadership roles. Glass walls may prevent both male and female candidates from taking on the roles, within teams, that they are truly best suited for—particularly in a glass escalator situation, where men may be pushed up the ladder to manage and lead the mission or the community, even if they prefer to hold back on such positions and even if the women among them would like to take these positions on.[14]

There is one way in which women may have opportunities for prestigious leadership roles. In a series of recent studies, social psychology professors Michelle Ryan and Alex Haslam showed that women were more likely to be appointed to authority and leadership roles when their organizations were in a crisis and at risk than when those organizations were experiencing periods of success. They named this phenomenon the glass cliff, referring to the risk of being appointed to a leadership position when the chance of failure is highest. Their findings suggest that, surprisingly, women are perceived as more suited than men to manage organizations that are in stressful situations—that people tend to "think crisis—think female" when appointing managers for failing companies.[15] Evidently, women's stereotypical qualities of being understanding, intuitive, and creative are viewed as particularly appropriate for crisis management in companies that are in precarious circumstances.

Of course, the risk of personal and organizational failure is high for managers who accept such positions. In these circumstances the co-occurrence of two relatively rare events—the appointment of a woman leader (e.g., director, manager) and continuing poor organizational outcomes—is likely to lead to such events being seen as meaningfully related, because this association fits with prevailing theories and ideologies. In this way, compared to men, women who assume leadership positions may be differentially exposed to criticism and in greater danger of being apportioned blame for negative outcomes, even those outcomes were initiated well before they assumed their new roles—a situation that was referred to as the glass cliff. Still, perhaps the riskier nature of Mars One's mission means women will have more opportunities to take on important leadership roles within it.

> "This project is not only about dreams and goals, but about the consequences for the future. It resonates in all of us to strive for something more, to participate in a legacy that will both provide progress and hope for future generations. Whether or not this project succeeds, it has done so much already in sparking interest."
> —Mars One applicant

Together, this suggests that women are more likely to find themselves under glass ceilings, between glass walls, and on the verge of a slippery glass cliff, while men may find themselves going up a glass escalator. All these recognized gendered dynamics are likely to limit women's ability to get to space by joining the Mars One leadership team, and to limit the ability of women on Mars to hold influential roles that are based on their character and abilities and not on their gender affiliation.

Encouragingly, in recent years the theory and practice of leadership and management have undergone a noticeable change. Industrial-era models have been superseded by newer ones considered more appropriate to the knowledge-intensive realities of today's workplace. These new, more stereotypically female models stress relational "postheroic" leadership. Many different labels are used to describe this new model of relational leadership, including such leadership forms as: distributed, distributive, benevolent, shared, quiet, connective, inclusive, humble leadership, postheroic, and complexity leadership. Although there are important distinctions, all share an emphasis on more egalitarian and collaborative, less hierarchical interactions and recognize the importance of relationships in fostering positive personal and organizational outcomes.

According to Joyce Fletcher, a researcher from The Center for Gender in Organizations at Simmons College, these relational or postheroic models of leadership have three characteristics that distinguish them from more traditionally individualistic models. First, leadership is shared and distributed rather than enacted by a single person. Second, leadership is a social process, in which human interactions are key. Third, leadership should result in learning, growth, and well-being for the people involved as well as for the organization.[16] If the Mars One community adopts such novel perceptions of leadership models, men and women will be more likely to share authority and contribute to the formation of a more egalitarian community on Mars.

Together Is Better: Gender-Integrated Teams

Bazooka brand chewing gum used to include an insertion with different jokes and predictions about the future. One of the favorite notes about the future for both boys and girls was: "By the age of 21 you will reach the moon." The promise of flying to the moon, in a message wrapped around a piece of pink chewing gum, was inspiring and delightful to many. However as they grew up, it became evident that the girls were less likely to "reach the moon" and space in general.

Given the limited number of women in space, teams in spaceships are more likely to be single-sex groups or employ a solo woman. This may not enable the best performance outcomes of spaceship members.

Many recent studies have indicated that gender-diverse teams may outperform teams that are not diverse. One such recent study, by professors Anita Woolley and Thomas Malone, showed that there is little correlation between a group's collective intelligence and the IQs of its individual members, but when a group is more diverse and includes a higher proportion of women, its collective intelligence rises. The research team gave subjects aged eighteen to sixty standard intelligence tests and assigned them randomly to teams. Each team was asked to complete several tasks—including brainstorming, decision-making, and visual puzzles—and to solve one complex problem. Teams were given intelligence scores based on their performance. Surprisingly, the findings showed that the teams that had members with higher IQs didn't earn much higher IQ scores, but diverse teams that included more women did.[17]

"Having different points of view when resolving problems is something very important in a team. On the Mars One astronaut team, we will be all in this together, and if one of us fails, we all fail."
—Mars One applicant

This only reinforces long-standing findings showing that groups that bring together individuals from multiple identity groups may outperform more homogeneous groups because they ordinarily include members with different ways of representing and solving problems.[18] Such groups may even be more successful than groups composed of individuals selected exclusively for their high ability because the best solutions to complex problems generally result from the application of different tools and abilities. The challenge in such identity-diverse groups is to leverage this potential by reducing the communication barriers, conflicts, and lack of mutual respect that can develop.[19]

A longitudinal study on research productivity, conducted on more than five hundred scientists and engineers in sixty work teams across a variety of science and engineering disciplines, showed similar findings. Examining differences in how men and women in science and engineering teams evaluate their colleagues' expertise and how this affects team performance showed that a team's overall gender composition predicts how well women's expertise is used within the team, and that gender-integrated teams with a higher proportion of highly educated women are more productive.[20] The teams studied, being comprised of scientists and engineers, are somewhat similar to the Mars One astronauts' teams in characteristics. An advantage for gender-diverse teams was also found when looking at creativity and profit. For example, some studies show that diversity in education and gender is likely to contribute to the introduction of an innovation. A field experiment conducted to estimate the impact that the share of women in business teams has on performance, based on venture teams consisting of undergraduate students in business studies, showed that teams with an equal gender mix perform better than male-dominated teams in terms of sales and profits.[21] Similarly, studies on directors in firms showed that female directors have significant impacts on board inputs and firm outcomes. In a sample of US firms, female directors had better attendance records than male directors, and male directors had fewer attendance problems the more gender-diverse the board was. CEO turnover is more sensitive to stock performance, and directors receive more equity-based compensation in firms with more gender-diverse boards.[22]

"Reading about the mission to Mars gave me a different perspective. It opened my eyes to the fact that the Mars One mission is essential to create change and bring progress to humanity."

—Gabrielle Ybanez,
Mars One applicant

All these studies suggest that gender-diverse teams of astronauts on the Mars One mission may outperform gender-homogenous teams.

Recent field studies in organizations also support this claim by showing that the ability to blend stereotypical feminine and masculine characteristics in an individual also affects outcomes.[23] For example, men who were encouraged to add feminine characteristics to their masculine style, even in masculine environments, were found to have better performance outcomes in terms of leadership effectiveness.[24] In an interesting case study of two offshore oil platforms, organizational researchers Robin Ely and Deborah Meyerson illustrated that establishing an organizational culture that is less masculine and that does not favor macho behaviors can improve organizational effectiveness and safety. By creating a culture that released men from societal pressures to demonstrate "manly" behaviors, such as proving how tough, capable, and bold they were (as was typical of men in other dangerous workplaces), the companies studied enabled men to keep a safer workplace and be more productive.[25]

Mars One's choice to enforce gender-diverse teams may also be of benefit when it comes to mission-related risk-taking. Many decisions involve choosing whether to risk something to try to obtain a potential reward, or navigate a safer course that will reduce both risk and potential reward. Overall, men have been found to seek more risk than women. For instance, in the United States, single men risk more of their assets and wealth in comparison to single women. A recent study has shown that acute stress amplifies sex differences in risk-seeking, such that men become even more risk-seeking and women more risk-avoidant. (The study's authors hypothesize that pressures of natural selection may have resulted in different biobehavioral responses to stress: In men, risk-taking during stress may have benefited them during competition for territory or other resources; the same tendency in mothers could have endangered the lives of dependent offspring.) This suggests that in stressful situations, such as the crew of the Mars One mission will face, a gender-diverse team would likely make more balanced decisions.

In other words, Mars One's choice to form gender-diverse teams is likely to be of benefit in a number of performance aspects, including creativity, safety, and the decision-making process. And if both women and men in these teams are able to show hybrid behaviors, it will likely result in even better performance.

Gender Talk: To Be Seen and Not Heard

A great deal of research has established that women are less likely than men to engage in aggressive or dominant behaviors. For example, in small-group contexts, women are less likely to engage in nonverbal displays, such as chin thrusts, or verbal displays, such as initiating speech.[26] Similarly, women are less likely than men to emerge as a leader,[27] initiate negotiations,[28] and behave aggressively.[29] Conversely, women are more likely than men to engage in low-power displays, such as smiling and maintaining eye contact when an interaction partner is speaking. Other academic research argues that women use less powerful speech: They tend to swear less, speak more politely, often weaken their statements, and use more tag questions and intensifiers. Women also tend to interrupt less than men do.

A few theories have been forwarded to explain these differences. One suggests that because men and women occupy different social roles, they behave in predictably different ways in line with these roles. Specifically, because women are more likely than men to be in nurturing roles (e.g., mother, caretaker), they may behave in ways that are more communal and less aggressive. Another focuses on men's greater levels of testosterone, which is associated with greater aggression and dominance.[30] Regardless of the cause, the result is that, in cultures where assertiveness is linked to confidence and capability as a leader, women are put at a disadvantage when interacting with others.

These differences in communication style suggest that this is another area where Mars One's teams would benefit from being wary, lest women's voices be less likely to be heard and valued. As we've seen, gender-diverse teams are significantly better positioned for success. Leaving out or devaluing women's voices limits such a team from taking full advantage of its assets and working to its full potential—a real danger in high-risk circumstances when a team should be drawing on all the talents of the group members to find the best solutions.

Gender Play: Rethinking Gender Relationships on Mars

Mars One aims to find ways not to replicate Earth's complex, limiting, and potentially harmful gendered dynamics on its mission and in the community that will be built on Mars. My recommendation as to how best to accomplish this is something I chose to call "gender play." Play can be defined in many ways and is complex in terms of its nature, purpose, and manifestation. Play has been defined in various ways as "a vacation from reality," "a purposeless activity," "fundamentally different from earnest activity," "an activity one is not obliged to do," and "voluntary intrinsically driven activity without a specific purpose that is done for its own sake and is associated with pleasure and enjoyment."

Stuart Brown, in his book *Play: How It Shapes the Brain, Opens the Imagination, and Invigorates the Soul*, contends that "at some point as we get older . . . we are made to feel guilty for playing. We are told that it is unproductive, a waste of time, even sinful . . . The play that remains is, like league sports, mostly very organized, rigid, and competitive. We strive to always be productive." But play, especially free play, has many benefits. "Play is a catalyst," Brown says, and its effects "can spread through our lives, usually making us more productive and happier in everything we do."[31] It involves freedom, surprise, and uncertainty, and enables loose and flexible associations between means and ends. Play is a strong engine for creativity and a way to think of directions that have not previously been thought of. Playful interactions, ones that are comprised of amusement, imagination, and make-believe, have the potential to contribute to the development of novel thinking about leadership and about gendered interpersonal and intrapersonal interactions, enabling Mars One colonists to dismantle, rethink, redefine, revise, reinvent, and construct a more egalitarian society on Mars One that will not be a mere replication of the existing gendered power structure on Earth, but rather enable both men and women, as well as other identity groups, to experience and live better lives together.

To do this, the Mars One missions need to construct "play spaces" for members of the mission and the community. Play spaces are safe physical and psychological settings that provide relief from the pressure of social validation and legitimize exploration. Such spaces allow players to try out new and untested identities, thoughts, and leadership behaviors and forms, and to engage fully in the learning cycle of feeling, reflection, thinking, and action. Constructing a culture of play and playfulness and enabling safe play spaces will allow a "laboratory" for innovation and creativity, which may result in a society that, apart from being more enriching and fun to be a part of, will allow for new forms and ways of enacting and constructing—or rather, deconstructing—gender and leadership. Through play, men and women can explore new ways of interacting with one another: reversed roles, exchanges of gendered roles, hybrid gendered roles, or taking turns in roles. They'll have a chance to explore new leadership dynamics and forms of postheroic leadership.

"A new life on Mars will bring forth the need to create a new social entity, detached from traditional human society. Given that the transformations needed to create a truly egalitarian society require a totally blank starting point, the new Martian society will be the best opportunity to create something not bound by any old social phenotype, but truly a social entity, created to serve the ones it consists of. Being able to participate in the birthing and rearing of this would be both an honor and a privilege."

—Mars One applicant

Perhaps the training and selection process itself can provide just such a safe play space, in that it will present challenges without real-world consequences in which candidates will be asked to experiment to find the most effective ways of working as a team. In such an environment the saying that "a child in play acts 'as though he were a head taller than himself'"[32] can be rephrased to, "A community at play, in space and on Mars, may act as an inspirational community that is 'a head taller' than the one on Earth," striving for a better world of new possibilities.

***Ronit Kark**, PhD, is a Tenured Senior Lecturer of organizational and leadership studies in the Department of Psychology at Bar-Ilan University in Israel. Dr. Kark received her PhD from the Hebrew University of Jerusalem, and completed her postdoctoral studies at the University of Michigan, Ann Arbor, USA. She is an affiliate scholar at the Centre for Gender in Organizations (CGO), Simmons Business School, Boston. Her work has been published in leading journals and presented at conferences around the world. Dr. Kark is the founder and first director of the graduate program "Gender in the Field: Linking Theory and Practice" at Bar-Ilan University. She received the Loreal-Rekanati Prize for the Study of Women and Management in Israel. In 2005 she was awarded the Best Paper Prize at the International Leadership Association (ILA) and the Academy of Management 2012 Award for the Scholarly Contributions to Educational Practice Advancing Women in Leadership. She works as an organizational consultant for leading organizations in different sectors (private, NGOs, and the public sector).* *She would like to thank Rachel Zalta, a graduate student in Social-Organizational Psychology at Bar-Ilan University for her valuable research assistance on this chapter, and Ronit Waiesmel-Manor and Leah Wilson for their helpful feedback.*

WITH THE WHOLE WORLD WATCHING

Imagine being interviewed for a job or trying to solve a problem with people you've just met, knowing that cameras are broadcasting each word you say and that an audience can see each grimace you make! The selection and training for the Mars One mission will indeed be very different from similar activities carried out for other traditional space missions: The last phases of narrowing down the short list of one hundred candidates (as well as much of the training process) will be filmed and shared with interested audiences across the globe. What impact will continuous filming have on the candidates? What is the desired impact on the audiences? In fact, why broadcast at all? And who are these candidates?

Before answering these questions, let's take a closer look at Mars One's plans for this filming process. There has been much speculation about what the filming will be like, and it has commonly been assumed that the selection and training of the candidates will be treated as in current reality shows. This assumption cannot be further from the truth.

In fact, the filming and the show that results will comprise a totally new format. First, the goal of recording the selection phases and training process will be to assess candidates' honesty, knowledge, integrity, and ability to work as a team. Inappropriate behavior (e.g., verbal abuse and discrimination, which is commonly courted for dramatic impact in many current reality-show formats) will lead to immediate elimination from the selection and training process. The drama will most likely come from competing in challenges and the personal conflicts that undoubtedly will arise at any stage and need to be worked through.

Story lines will not be generated ahead of time, and scenes will not be staged or restaged for the cameras. There will be no confession chambers used during filming or in the days and months afterward. Participants will not be coached in what to say or how to behave. Moreover, there will be no leading questions or attempts to elicit specific responses, and no intent to humiliate or exploit participants—though direct questioning, of individuals or in group debriefs, will take place in order to elicit participants' thoughts, judge their mind-sets, and reveal their true characters.

Because healthy and effective teamwork is essential for the first Mars settlers, the challenges will always be addressed to the team—the team as a whole wins or loses. Participants will rank the order, at the end of each selection day, of the three teammates with whom they want to continue. Mars One Selection Committee members will then, together, decide on any members who should be eliminated, based on what is good for an optimal team. (However, in some cases, the participants who are proposed for elimination will have the chance to defend themselves and argue their cases.) Any groups that win a challenge will not lose a member in that specific challenge round.

The reward for the winners will not be money, but further education and training, and possible settlement on Mars. Ultimately, the goal of this exercise is to find the right settlers for the Mars One mission.

Still, the process will inevitably have some things in common with the traditional reality-show format. In this section, Cindy Chiang Halvorsen, *winner with her husband, Ernie, of Season 19 of* The Amazing Race, *recounts in her essay "The (Amazing) Reality of Being on TV" their experiences on that show and what it was like to approach tasks and work as a team while being filmed almost every inch of the way. Their experience highlights some of the related challenges the Mars One candidates must also be prepared to face.*

Given some of the difficulties Halvorsen notes as innate to filming, why film the process at all? Dr. James Kass, *one of this book's editors and member of the Mars One Selection Committee, considers in his essay "How Filming Mars One Could Change the World" the desired impact of filming the crew selection and training process on viewers at home.*

Of course, keeping the audience entertained and increasing the number of viewers is the goal of all televised shows. But the Mars One Selection Committee members want to go about achieving these two goals differently than anything else currently on the air, and Dr. James Kass explains the ways Mars One hopes these episodes could affect viewers' lives beyond just entertainment, to help change attitudes toward the human conquest of space on the kind of mass scale television makes possible.

Finally, the final one hundred Mars One candidates themselves—the ones who will be in front of the camera—are given a chance to speak, in "Inside the Minds of the Mars 100." Who are these people that are willing and eager for a one-way ticket to Mars? Are they crazy? Are they very unusual—would they stand out in a crowd, or simply pass as people who would never consider venturing out to such an adventure?

The last chapter of this section brings the views of the candidates to us in selected answers to the questions posed to them as part of the selection process. Their answers reveal a mosaic of characters and personalities that help us to understand who it is we will be watching as the next phases of this project are broadcast to the world.

—The Editors

What is the impact on participants of filming the crew selection and training process?

THE (AMAZING) REALITY OF BEING ON TV

CINDY CHIANG HALVORSEN

Thirty-five thousand miles, twenty cities, nine countries, and four continents traveled. My name is Cindy and together with my husband, Ernie, we're the winners of *The Amazing Race, Season 19*. *The Amazing Race* is a reality-TV show where eleven teams compete in a race around the world. The teams, each composed of two participants with a preexisting relationship, travel to various locations in the world, engage in a series of mental and physical challenges known as Detours and Roadblocks, and attempt to be the first team to complete the route and arrive at the finish line.

What does this have to do with Mars One and its mission to Mars? Having completed the most extreme race around the globe, Ernie and I learned a thing or two about the importance of working together as one team to conquer the unknown. And after being filmed day and night for more than three weeks, we also know a little something about what it's like to do those things with millions of people watching.

The astronauts vying to fly to Mars won't just have to worry about the challenges awaiting them during colonization. First, they will face rigorous training—their own mental and physical challenges—to prove themselves capable of the mission. This whole process will be caught on film.

Stronger as a Team

Unlike Ernie and me, candidates for the Mars One crew-selection process won't know one another beforehand. Although team members will be able to choose one another based on certain criteria—each team must be half female, half male, and as international as possible—they will not have preexisting relationships, so will have to quickly figure out how to work together and navigate one another's personalities.

For *The Amazing Race*, Ernie and I applied to compete as a team. We had the luxury of already knowing each other intimately and how to work together. Early on in our relationship, however, we had to learn how to navigate our differences—and our differences were striking. We may as well have grown up on different planets! Ernie is a homegrown all-American boy from a working-class neighborhood in Chicago who spent most of his childhood on a baseball diamond or playing hockey in the city's alleyways. In college, he was always scheming up pranks and throwing parties. Ernie's attitude is risk-taking, sometimes reckless; he has a "take life as it comes" mentality.

> "I believe all things are possible when we are unified, and I cannot think of anything more unifying for this planet than putting one of us on another world."
>
> —Mars One applicant

I was born to Chinese immigrants who valued education first; both parents are professors. I grew up with a busy schedule with traditional school, Chinese school, art class, martial arts, and piano lessons. During college, I was an active campus leader and graduated top of my class. I later earned my MBA from Northwestern University while still working full-time. When I have a goal in mind, I set out a plan to achieve it, which means I can be extremely intense.

We fascinate each other and our differences make us stronger as one team. I'm a planner; Ernie makes decisions on the fly. Where I approach problem-solving logically, Ernie comes up with more creative solutions. When I get too serious, Ernie knows how to lighten the mood. Sometimes we still clash when Ernie doesn't understand why I'm so intense, and I can get frustrated when I feel he isn't being serious enough. These differences apparently made us interesting reality-TV contestants. In our audition tape, we played up our "opposites attract" relationship and caught the attention of the casting director.

We learned to embrace our diversity and use it as an advantage in the game, and the final two episodes of the race are a great example of how our differences led to victory. In Panama, we couldn't figure out the last clue identifying the location of the last Pit Stop. As the risk-taker, Ernie pushed us to get into the taxi even though we had no idea where we were going. Luckily, the taxi driver had gotten a tip from another team and took us to the right place, securing our spot in the final three. Then, during the last leg in Atlanta, I had to trace the course we had raced on a giant world map while being suspended fifty feet in the air. This task played to my strengths because I, the planner and prepper, had already memorized the racecourse I had drawn on a map during filming. I quickly accomplished the task and off we went to the finish line to claim the $1 million prize!

Neither of us could have successfully completed the race alone. While the Mars One candidates will be facing very different tests, each team's success will depend just as much on their combined skills . . . and their willingness to learn about and trust each individual's unique abilities and strengths.

Preparing and Training for the Race

The men and women vying for spots on Mars One will be getting to know one another the same way the selection committee and viewers at home will come to know them: through the selection and training process. Because this process will be filmed, the Mars One documentary is more similar to reality shows such as *Survivor, Big Brother*, or *The Real World*, where strangers are thrown together and need to work together (or at minimum, coexist), than it is to *The Amazing Race*. For us, winning was the end goal; for the Mars One candidates, "winning" is only the first step.

Still, Ernie and I completed our own training in preparation to compete. That preparation, in which we got to know each other and ourselves better, proved to be our greatest strength.

Even before we knew that we'd scored one of the coveted spots on *The Amazing Race*, we laid out a comprehensive six-month training plan to get us physically and mentally in shape for the tough road ahead. We have been huge fans of the show for many seasons and always imagined ourselves competing alongside the contestants on TV by picking the Detours and deciding who would do the Roadblocks. Our main strategy was to be prepared for any situation, and we knew that the challenges would test a variety of mental and physical skills.

We completed an extensive analysis of our own strengths and weaknesses, so that when it came down to deciding who would tackle a Roadblock or which Detour we should take, we were prepared to make an immediate decision rather than having to stress over it during an already stressful situation. Ernie excels at sports and physical activities; my strength lies in the mental challenges, like memorization and puzzles. Having a solid understanding of these abilities helped us to develop our game plan and act on it.

While on the racecourse, every time we were faced with a decision about who should complete the challenge ahead, we reflected back on our individual strengths assessment. It was a quick and easy decision for Ernie to take on physical challenges, like moving hundred-pound bags of tobacco in Malawi, or racing a Mustang on the proving range in Copenhagen. My listening and memorization skills made me the obvious pick to memorize an old Confucius proverb and recite a Hans Christian Andersen poem. Speedy decision-making is critical to succeeding in the race; every second is valuable and could mean the difference between winning and losing the race.

> "The Mars One endeavor is significantly greater than anything I could hope to accomplish on my own. This is an adventure set to define a lifetime. It offers challenges that have not been undertaken before, and the opportunity to define the adversary."
>
> —Mars One applicant

We kicked our physical training into high gear. We were out running with backpacks, rock climbing, kayaking, swimming, and weight training. Mental training was also important. We studied maps and travel books. Knowing that verbal communication skills would be extremely important and, honestly, not one of our strongest areas, we specifically set aside time to work on it. This was definitely the toughest part of our training, despite having known each other for years. We bought a Lego kit with instructions on how to build a truck. Ernie took the Lego pieces and I took the instructions. Sitting back-to-back, we had to build the truck, with him never seeing the instructions and me never seeing the pieces. We learned a lot about communication and even more about patience. All this preparation allowed our relationship to evolve and strengthen before *The Amazing Race* was filmed.

Caught on Camera

Our training process happened off camera, so all of our "learning moments" and disagreements were thankfully untelevised. The Mars One contestants will be working through all of this—getting to know one another, learning one another's abilities, and discovering how best to communicate with one another in stressful situations—while being filmed and competing against other teams. In addition, the Mars One team members will be learning valuable survival skills, such as repairing their habitat and rover, medical procedures, and how to grow their own food in this new environment, all while the cameras are rolling. It should make for some interesting television!

One of the common questions Ernie and I get asked when people first learn about our being on the *Race* is, "What's it like to be filmed all day?" First, let me explain how filming works on *The Amazing Race*. Although we are depicted as a team of two, we actually ran the race as a team of four: Each team of two racers is paired with one soundperson and one cameraperson. At any given time, all four of us had to be within twenty feet of one another to ensure everything was captured on film and audio. There were additional location cameras and sound teams for the task areas, key route points, and the Pit Stop for each leg of the race.

Although for much of the race we felt like regular people competing on an amazing course, the actual logistics of filming impacted how we behaved. We always had to consider the extra two people along (camera and sound), because the four of us had to stick together. These two additional team members had a different priority than we did—they needed to get great footage—but we depended on them to move quickly through the race. For example, while we were racing in Indonesia, I rappelled deep into a cave to get the next clue. Our cameraman followed me down but then was held back by producers to get more footage inside the cave. We had to wait twenty minutes until he emerged before we could continue on. (Twenty minutes could mean the difference between being first or last!) Delays like this happened to many teams over the years.

Furthermore, there's the equipment you have to accommodate. Because *The Amazing Race* travels through various countries, all of the filming equipment must pass through cumbersome customs processes, which can certainly delay teams competing in the race. Sometimes the camera's battery pack runs out of juice and you have to stop to change it. Other times, perhaps the body mic isn't working properly and you have to stop to readjust it. If there was a water challenge, we'd stop to switch the mic equipment to a waterproof mic. The cameras always got foggy when going from an air-conditioned interior to a hot and humid exterior, so we had to wait to defog the lens, or do without the AC in a hot taxi. These are all standard procedures and typical events when you're filming a great reality show.

While the Mars One candidates will largely be anchored in one place and won't have border customs or air-conditioning to worry about (though a broken heater could mean the difference between life or death!), they presumably won't be able to avoid the reality of the sound and video equipment, which can make an already manufactured situation feel even less real. At first, it felt like we and the other *Amazing Race* contestants were running from one Hollywood movie set to another. It was most noticeable at our very first Pit Stop in Taiwan. When we arrived at the check-in mat where host Phil greeted us, we were surprised when an entire entourage swooped in with bounce boards for light reflection, more cameras, and sound booms. It didn't quite feel like we had actually raced from Los Angeles to Taipei despite long plane flights and nights spent in airports.

> "I love Earth, I love its people, and I love my life here. I want to be a part of Mars One not because I want to leave Earth, but because I want to be one of those that help establish a toehold on Mars for all of humanity. I believe this mission is so immensely important that leaving Earth will be easy, because I would be going *for* Earth."
>
> —Mars One applicant

At the start of filming, there were eleven teams, essentially forty-four people, running around each area searching for clues or completing a task. Add to that producers, security personnel, location cameras, and sound people, and it would be hard to miss that something is being filmed. You're just always conscious there is a camera in your face. It was very important for several contestants in our season to always appear camera-ready, with perfect hair and makeup (some even wore fake eyelashes), and they often gave up time to eat or sleep in order to groom themselves.

Most people don't act as they naturally would when they are hyperaware of being recorded on camera. Some people are overly dramatic; they know that if they overreact, they'll get more visibility once the show airs on national TV. These extreme personalities are what make most reality-TV shows so entertaining; after all, drama sells! (Mars One candidates for crew selection may also exaggerate their behavior to try to get airtime. Also, as with any reality show, its producers will be trying to attract advertisers, and more dramatic characters can lead to higher sponsorship potential. Still, we hope that Mars One participants will focus on learning survival skills; their stakes are higher than the $1 million prize at the end of *The Amazing Race*.)

> "I believe that the key to the continuing progress of the human race is in having more questions to ask than there are answers to. That is the drumbeat that we as humans march to. By being part of a team that is selected to go to Mars, I am hoping to be in the unique position of opening the doors to many more questions, and along the way, finding some answers, perhaps to my question, and bolster humanity's understanding just that tiny bit more."
>
> —Mars One applicant

Displaying motive-driven changes in behavior is the most obvious impact the camera can have on participants. For me, being followed by a camera crew made me self-conscious, like someone was constantly judging what I was doing. It was hard to be totally natural knowing millions of viewers would watch everything I did and said. At the same time, I was somewhat intentionally creating a TV character that fit the archetype that originally attracted the casting directors—the high-strung, type A overachiever. Everything was about winning, about being the fastest, the smartest, or the strongest. Although there are definitely elements of the real Cindy in there, ultimately, the character on the TV show was just that—a character with exaggerated traits of a real person.

The producers who conducted the interviews for *The Amazing Race* asked clever questions that would elicit the responses they needed to tell a great story. It's their job to create a show that millions of viewers find riveting. As a contestant, you want to provide great content and help them do their job well, which is further fueled by the desire to get more airtime. When asked a question during interviews, you're coached to respond by restating the question first and then providing your answer. Producers might ask: "Are you the strongest team still in the race?" So your response would begin: "We are the strongest team still in the race because . . ." Then the next question might be, "Who is the weakest team?" Your response then allows them to tell a story of team rivalries.

Ernie reacted to being filmed a bit differently than me. He was much more politically correct with his answers than he typically would be for fear of eliciting bad public opinion. As he put it, he didn't want to give his friends any reason to make fun of him when they watched the show. When you're on camera, there's an internal monologue that goes through your head; you're second-guessing everything you say simply because you don't want to be caught saying the wrong thing. Ernie would have responded to that second interview question with, "I respect all the teams and they're great competitors. Hopefully, in the end, we still come out victorious," whereas off camera, he would have normally given a highly emotional and animated commentary. (His friends didn't make fun of him when we got back home, but while his politically correct answers were eloquent, they didn't give the producers much help in creating an entertaining TV show. Let's just say, in the end, my interviews got more airtime.)

After the first few days of filming, you begin to settle into a rhythm of how you want to act in front of cameras and ease into that character you've subconsciously—or consciously—chosen to create.

Producers of a reality-TV show about Mars One plan to approach capturing thoughts, feelings, and emotions in a different way than the producers of *The Amazing Race* did. On *The Amazing Race*, we had interviews before leaving each starting point, on the fly during the race, and after checking in to the pit stop. The candidates for the Mars One mission will be encouraged to express their natural thoughts and feelings out loud during all of filming. When they are given specific questions, the psychologist asking them will also be filmed and shown on-screen, which should prevent leading questions.

Creating an Entertaining TV Show

It's not just the way people being filmed behave that makes a reality show different than reality; it's also the influence of the show itself. In our experience with *The Amazing Race*, producing an entertaining TV show was a clear priority over running the race. During filming, we were often reminded that this was a TV show first and a competitive race second. The producers want to uphold the integrity of the game, per the written rules, but it is a TV show and it doesn't matter to them who wins or loses. The show must be entertaining from the start to the finish line, whereas naming the winner is only interesting for the last few minutes of the season. In the end, the producers want to get the clips they've planned ahead for and they want to tell the story the way they've scripted it. You're merely the participant that's lucky enough to get on film.

(And I do mean lucky! Very few people get to see the world in this exhilarating way, one that also teaches teamwork skills and leadership lessons. Ernie and I are so grateful to have shared this experience together and become a stronger couple. My type A personality means I often want everything to go as I've planned. But through the process of the race, I learned to give up some of my high-strung attitude to be a better partner to Ernie. I shifted my priorities from wanting to always win and be the best to living a more fulfilling, purpose-driven life and striving to be a positive influence for others.)

While the *Race* seems like a spontaneous show, competition reality shows, in general, are expertly crafted through casting, editing, and controlling the filming environment. Contestants are selected to fit a character, and editing can accentuate these personalities, since time restraints require hours' or even days' worth of video to be cut down to short segments. It is the producers who control where the action takes place and, when challenges are involved, what those challenges are; in our case they controlled the locations and tasks to be accomplished. The casting process helps producers to anticipate the reactions of the participants to those locations and tasks—and the more extreme those reactions are, the better.

The producers wanted to get the widest variety of emotions to improve the show for viewers. People acting their best is not what makes entertaining TV; it's the most animated parts of their personalities that does that. So what did they do to ignite extreme emotions? Starve us. Deprive us of sleep. Keep us dirty. Induce anxiety through the fear of the unknown. Can you think clearly when you're tired, your stomach is growling, and you can smell someone else's feet because they haven't showered in days? (They could get us drunk, like on some of the other reality shows—ahem, *The Bachelor*—but that'd be too dangerous on *The Amazing Race*; imagine tightroping between two thirty-story buildings while tipsy!) Then they took all those raw emotions and edited them down into an hour of suspense and drama.

This is where our experience and the experience of the candidates for the Mars One crew will likely differ most. Mars One's show aims to be as educational as it is entertaining; its content is less about emotional drama and manufactured suspense and more about the reality of space travel. Story editors will be focused on teaching the audience about Mars and the challenges these astronauts will face as they prepare for their odyssey. Mars One will no doubt still test teams' ability to work well under pressure, with little sleep and in less than optimal conditions, but the endgame will be to prepare them for what they might find when they land on Mars, not to entertain the viewers. Teams will be incentivized to work together rather than competing with one another to get ahead—alliances built not on trying to further their own team's chances (or not only on trying to further their own team's chances), as they are in most reality shows, but on a shared belief in the overall mission: colonizing Mars.

It may be unavoidable that contestants will be influenced by the film crew or behave differently than they normally would because they feel like they're "on-set." Let's just hope that Mars One's "set" helps prepare these astronauts to be the first humans to live on Mars. Because, in the end, our thirty-five thousand miles around the globe pales in comparison to their thirty-five million miles to Mars.

Cindy Chiang Halvorsen *is lucky wife to Ernie and mama to their son, Maverick. She loves her career as an award-winning brand marketer and business strategist. A portion of Ernie and Cindy's* Amazing Race *prize money is invested in social impact start-ups that create sustainable employment for people in developing markets.*

What is the desired impact on viewers of filming the crew selection and training process?

HOW FILMING MARS ONE COULD CHANGE THE WORLD

JAMES R. KASS

As a pupil in school in the 1950s, I remember the animated discussions we had about a new, exciting, and adventurous field—spaceflight—that was just opening up. We had no TV or internet, but we were nevertheless well informed. The Soviets had just beaten the Americans to launch Sputnik, the first satellite, into space. Then the Soviets succeeded in beating the Americans once more by orbiting Yuri Gagarin, the first man in space. Next, the Soviets claimed to have achieved the first spacewalk; but this time we had doubts—there had been accounts of some telltale marks on the released pictures that looked suspiciously like strings. Could it be that the spacewalk had really been a simulation, maybe underwater? In fact, how were we to believe that Gagarin really went into space—maybe that was a hoax, too?

I went to the local library and borrowed every single book about space, whether about astronomy or how to build a rocket, and read them all through (not understanding quite all I read). A dream was developing. I did not know exactly what form it would take, but I was certain: I wanted to work in this new and exciting domain. Science fiction had become reality. I would enter university and study all that I had to in order to achieve this goal.

Fast-forward more than twenty years into the future, 1983. I am sitting in a windowless room at Johnson Space Center Houston in Texas, watching a chart recorder. I am wearing a headset with a microphone in front of a panel of communications channels. The crew I recently trained have been launched into orbit on the space shuttle with the first European-built Spacelab. The first scientist-astronauts dedicated to operating Spacelab were on board and running our experiment, and I was in communication with them. I had realized my dream (and accepted that the Soviets had indeed achieved the first man in space and spacewalk!).

When I ask professional colleagues of my generation what first inspired them to invest many precious years of hard work to complete studies in challenging and difficult subjects, the answer is often the same: the excitement of the early days of the space race. But that was more than fifty years ago, in a very different world. Can the filming of the selection of Mars One candidates and their subsequent unique training bring back that inspiration again? I think yes, it well could!

Bringing such scenes to a wide audience could break up the feeling of stasis that has afflicted national space programs since the 1970s. In the decade following Sputnik, the United States alone developed four new spacecraft—the Mercury capsule, the Gemini spacecraft, the Apollo Command and Service Module, and the Lunar Module, *Eagle*. And each mission was groundbreaking in some way.

But today, every ninety minutes, the International Space Station completes yet another orbit, just as it has done since the first module was launched in 1998. It was preceded by the Soviet/Russian space station, Mir, which itself was preceded by the Soviet Salyuts and the US Skylab; these stations have been orbiting Earth over and over again during a period of more than thirty years.

How could we not expect the public to get rather bored with all these thousands of rather low orbits? Is there much going on to inspire the young people of today? I have worked with and at the European Space Agency's largest technical and research center (ESTEC) in the Netherlands for more than thirty years—and yet, when I tell this to nearby residents, I usually get a blank look. They have never heard of the place, there is so little interest!

Are we then surprised that devoting time and effort for many years to study a difficult technical or scientific field is the path taken by a progressively diminishing number? And that, despite relatively high unemployment, there is a dearth of such trained individuals to fill vacant posts?

But a new and exciting horizon is now opening up with the likes of a one-way trip to Mars (Mars One), a Mars flyby (Dennis Tito), and new spaceships built by private companies (SpaceX, Virgin Galactic). If the goals of these projects are met, maybe we shall have a new generation of inspired young people ready to slog it through difficult studies for many years to achieve new dreams. The hope of Mars One is that following its progress, and documenting science fiction as it unfolds into reality, will play its part in inspiring the next generation.

Will This Show Be Just Like the Other Reality Shows That Have Filled Television Screens in Recent Years?

Since I became an advisor to Mars One, I have often been confronted with questions posed by journalists and members of the general public about the filming of the selection process. Will the subjects be eliminated one by one as the audience submits their votes? Will each step of the selection process be based on how well one can perform to an audience, or how cunningly one defeats the "opponents"? My hope is that this will certainly *not* be the case.

So what will it be like instead? What impact will it have on the audience?

I envisage that these events will have both entertainment and pedagogical value. Just as with successful films that become classics, to be watched over and over again by young and old, I imagine that the Mars One selection video will prove to be exciting, informative, entertaining—even educational—and will provide a real portrayal of human nature with its wonders and flaws.

As the multicultural crew, also of varying ages, tackles the challenges posed to them, the audience will watch scenarios unfold, probably in a very different manner for each team. How will the group function as a team rather than as a group of individuals? With the different cultural mixes present in each team, one can expect a great diversity in the stories that arise.

A multicultural crew is not always a happy mix and can result in unplanned challenges.

I recall a story told to me by a former Soviet cosmonaut who served on the MIR station during the Shuttle–Mir Program. The project, sometimes called "Phase One," was intended to allow the United States to learn from Russian experience with long-duration spaceflight and to foster a spirit of cooperation between the two nations and their space agencies. The project helped to prepare the way for further cooperative space ventures—specifically, "Phase Two" of the joint project, the construction of the International Space Station. Each crewman was allowed a limited amount of personal goods to take with him during his long stay in orbit. An example of an unplanned challenge occurred during one of these cooperative missions: One cosmonaut took with him a special Russian sausage (*kolbasa*—not to be confused with the small sausages of English or German type) homemade by his grandmother. He consulted with his fellow Russian crewman whether or not to share this precious sausage with the American who was currently on board the station; this would mean less for each of them. The difficult decision was reached—to share—and each crewman received his carefully allotted third of the sausage. The Russian crew slowly ate small mouthfuls of the delicious sausage, realizing this would be the last such treat for many months. The American quickly consumed his share, and then, as the Russians waited expectantly for the sounds of appreciation and delight, he said lackadaisically, "Well, not bad, but certainly not as good as an American hot dog." I understood that communication and collaboration among this international team sharply deteriorated after this unfortunate incident.

The selection and training processes will contain challenges planned by the Mars One team as well as those everyday challenges arising spontaneously, when people are together, as will happen also on a mission to Mars. Some, as in the situation illustrated above, are parts of everyday life and normally challenges of a minor nature, but which could have more serious long-term repercussions.

Living together in a closed and isolated system will undoubtedly expose the team members to those natural challenges of living and being in constant close proximity to one another. Even natural bodily functions will not be exempt from unwanted sharing and could lead to unexpected and undesirable consequences.

> "It is not often that one gets a chance to make such a significant contribution to humanity. The mere knowledge of this fact, the importance of what I would be part of, would always keep me focused on the path lying ahead and not the one I left behind."
>
> —Mars One applicant

I recall, as part of preparing experiments for spaceflight and training the crew to operate them, we often simulated such operations in so-called parabolic-flight campaigns in aircraft flying reduced-gravity profiles, such as the NASA KC-135 or the Russian Ilyushin-71. We had to undergo rapid-decompression training, which some of us did with the help of the US Air Force in Wiesbaden, Germany. Natural body functions, such as passing gas, were part of the training; if one did not succumb to such needs during decompression, it could cause untold physiological damage to the body under decompression. One participant tried valiantly to suppress passing gas because there was a young lady sitting next to him—the result was almost fatal! In a closed small chamber where several adults are living together one can expect body odors and "outgassing" are no longer personal experiences, but can be involuntarily shared. I recall one ground engineer working on the Biorack facility during an IML-Spacelab mission suddenly notice an unusual and unexpected signal on the ground display of the space facility. After much fruitless investigation it was concluded that an astronaut standing next to the facility on Spacelab had passed gas! Almost nothing could be kept personal: This will also be the case for the filming of the selection and training processes—it will be open to all the public viewers!

Although all participants will share a common language (albeit not a common mother tongue), communication can still lead to misunderstandings and awkward situations between people of different national backgrounds and cultures. I recall some incidents during the many years I spent as an expat in various countries that taught me that even mastering a language is not enough to avoid or preclude embarrassing or pitiful situations from occurring. The same words spoken by different people, of different genders, nationality, or circumstances, can have very different meanings.

An old Russian saying goes, "When a politician says YES, he means MAYBE; when he says MAYBE, he means NO; and when he says NO, he is not a politician. For a lady it is different: When she says NO, she means MAYBE; when she says MAYBE, she means YES; and when she says YES, she is no lady!" Although this particular example is not always true today, there are still many others, not quite so amusing, which can result in gross misunderstanding when a spoken word or signal can mean different things for different cultures, languages, genders, religion, and so forth.

It turns out that eating, which can be a wonderfully sociable experience, can also be the cause of unhappy misunderstandings, despite being accompanied by a common language.

I recall the story an old man told me during my years in Egypt: When he was seven years old, this man attended a Sunday school run by American missionaries. One Sunday morning, the teacher brought a basket of apples and started distributing them to the young pupils. He was the first. When she offered him an apple, as was the custom in his country, he politely refused, expecting that she would offer a second time. The next pupils learned fast—and accepted immediately. He went home crying—he was the only one without an apple. Although they spoke the same language, the cultural meaning of "no, thank you" was rather different. In Egypt, one always has to refuse an offer at least once before accepting; this is because one is always bound to offer what one has as a sign of politeness, without the expectation that it will be accepted. So if you want to be sure that the offer is genuine, wait until it is repeated for the third time, then it is safe to accept. Knowledge of a common language means little without understanding the customs of a community.

Still on the subject of eating—I recall a painful incident my brother experienced when he visited an Egyptian classmate's house and was offered dinner, which is usually made up of several generous courses. The first course was *molukhiyah*, an ancient Egyptian culinary delight; for most expats, eating this mucilaginous concoction is worse than forcing a child to drink cod-liver oil. While no one was looking, he managed to empty the entire plate into a nearby plant, and looked forward to the second course, which was much more to his liking. When his classmate's mother saw the empty plate, she was delighted and instead of serving him the second course, filled his plate up again with the dreaded dish! What my brother did not know was that an empty dish told the host that he had not had enough and wanted more. Showing appreciation in different cultures is often demonstrated very differently, and can sometimes have the opposite effect than intended. This will inevitably happen during the early selection process for Mars One, which could be entertaining for the audience, as well as informative.

Although all crew will speak a common language, English, this does not mean that things will go smoothly—especially during the first tests, when they have not yet gotten to know one another's customs and habits well. The viewing audience will be able to observe these initial potentially embarrassing situations and the awkward learning process as the subjects handle unexpected scenarios, whether planned or naturally occurring.

I recall going to a conference in Austria with a few international colleagues from North America, Europe, and Southeast Asia. The conference was located near some beautiful skiing slopes and took place in the middle of winter. We asked our Asian colleague if he would like to go skiing with us, to which he politely replied in the affirmative, as he did to all our other questions about what we were planning to do. Rising up the mountain on the chairlift, we asked if he would like to accompany us to the very top (and most difficult) slope, to which he again nodded politely. When we arrived there and got ready to ski downhill, we were horrified to see that he could barely balance himself on skis, let alone ski! As was his local custom, he had tried hard to please us and agree to our suggestions—and as we had not bothered to really understand whether or not he could ski, we found ourselves in the challenging situation of trying to take him down the difficult slope, two of us trying to support him from each side, and nevertheless the whole company continuously and ignominiously falling over every few meters! We learned the difficult way about different customs of communication across cultures. With the wide variety of cultures and nationalities during the first challenging sessions of crew selection, the viewers may well be witness to many difficult and fascinating situations.

"Living life differently was always my dream and I am sure Mars is going to be the stepping-stone for that dream. I enjoy doing the things that others hardly think."
—Mars One applicant

Some individuals may relish being filmed, while others will wish to avoid the recording of their each and every move. Celebrities often have a love-hate relationship with the camera, as will possibly also the Mars One candidates. On my visits to Star City (the Russian cosmonaut training center), I often met Russian and international crew training for flight on the Mir station. On one visit, my Russian colleague pointed out to me an American astronaut who was training at Star City and about to fly to the station—he was walking past us. I raised my camera to take a picture of him and called out to him. He simply raised a sheaf of papers to block his head from my camera as he walked by. (He later apologized to me when, as a VIP visitor to Star City, I was inevitably formally introduced to him—he said that he thought I was simply one of those pesky tourists . . .)

I envisage that the selection and training processes will be a combination of varying types of activities, which potentially could all be filmed. These could include interviews, group dynamics exercises, and technical problems and various tests to demonstrate the propensity for skills, whether technical or psychosocial, necessary to survive on a mission to Mars.

The audience watching the process of testing and selection of those candidates will also learn about the valuable skills and characteristics that make a good candidate. In addition to being entertained, some will go away having learned more about these skills, how they are put into practice, and how important they are in life, on Earth as well as on any space mission. Some may even go away after each episode with a hope or goal to hone those skills themselves, as they begin to understand how such skills can help them not only to survive but to shine in their vocation and everyday activities. Many others, hopefully, will be inspired to become supporters of humanity's quest to explore outer space.

The individuals and the teams will be put through grueling questioning, challenges, and tests of their technical, social, and all manner of practical skills. Feedback to the teams will also be shared with the audience. As the individuals and teams go through the selection and training processes, they will learn a lot along the way—this will be a unique, and probably unplanned, form of learning not commonly available in regular institutions, and rarely aired and shared online and live.

There are only a handful of space agencies today that select and train astronauts for spaceflight. Although the prerequisites for selection and selection criteria are publicly available, none share with the public the most interesting final stages of how they eliminate those candidates who fulfill all the selection criteria, but who will not have a chance to fly. The Mars One selection process will be unique in that these last stages will be filmed and broadcast to the public, for all to watch and even provide feedback.

The requisite skills and characteristics of a Mars One space traveler will be very different from those today selected and trained to fly, for example, to the International Space Station. This reflects the evolution of requirements and expectations for astronauts, which has been happening since the first launch of humans into space. Those who had the "right stuff" for NASA's original group of astronauts were for the most part test pilots, strong lone heroes able to face the most dangerous circumstances single-handedly. In fact, most actual piloting in spaceflight has been automated and no longer requires particular piloting skills or heroic strength or prowess. Over the years, most astronaut candidates came to possess PhDs in engineering, physics, and other hard sciences. A flying license is no longer always mandatory. As Mason Peck points out in his chapter on the expected requisite skills of a Mars One astronaut, the candidate today will be far more in need of practical technical skills than high academic degrees. The average viewer, although possibly impressed by the PhD title, will be far more able to identify with and understand these practical skills. The audience will be more equipped to follow the physical solutions to concrete technical challenges facing potential crew members than they would to follow complex discipline-specific tests or even tiresome academic questioning. They will be able to learn from what they see—and hopefully try to apply these lessons with their home toolkit or at their local technical college. They will often be saying, "This is fascinating. I can try that myself," perhaps having never before thought such work could be carried out without special purpose, tools, or many years of study!

But technical skills alone, without psychosocial skills, are a nonstarter. What use is it to be able to repair a broken piece of equipment if the next moment one might be ready to harm a fellow crew member? The potential Mars settlers will certainly be tested and trained to face psychosocial challenges—indeed, to become their own psychotherapists. They will learn how to recognize potentially dangerous signals well in advance and take timely preventive or corrective action.

Both the selection process and training will take a holistic approach as to how to face and solve challenges: A team can be far more effective sharing their expertise than a group of individual experts unable to communicate effectively.

This theme is handled in greater detail in other chapters ("A World Waiting to Be Born" and "Culture and Communication"), but it suffices here to understand that the audiences will be exposed to various aspects of the dynamics and challenges of intra- and intergroup behavior; and the fact is that this is a common aspect of everyday life, whether at home, with our neighbours, friends, colleagues, those we like, those we hate, or those to whom we are indifferent. The challenges involved, their solutions, the art of leadership under different circumstances, the approach to problem-solving, and the skills to handle conflict are all part of the psychosocial toolkit that we have to hone in order to face simply living our lives on a daily basis. Watching the candidates face these psychosocial challenges, how they handle them, and how they learn from them will not only be interesting to watch, but should also be a learning experience for the audience.

The Filming of Mars One Selection and Training Activities as Basis for Educational Online Courseware

In the Information Age we are blessed with the availability of an abundance of online and downloadable courseware on almost any subject under the sun. We need no longer be hemmed in by a particular syllabus that is either too advanced or too basic, or a lecture series that goes too fast or too slow. Finding and following the right online courseware, which we can follow at our own pace and repeat segments as needed, is a wonderful luxury.

So, what has this got to do with filming of the Mars One training? The answer is simple. Some ten years of grueling training will be given to the future Mars settlers. This training is not likely to be highly erudite, delving deep into academic subjects such as we would normally expect in a university and that would only be of interest to a few scholars. Although mission-specific technology will obviously be involved, a lot of the training will be of a practical and general nature, such as how to perform some repair or handy work without the proper instruments, or invent a solution to an unforeseen problem, but with limited resources. This is not only easier than advanced academic subjects for almost everyone to follow, but is also useful in everyday life for any one of us. The broadcasts of such training could well become the basis for so-called Massive Open Online Courseware, or MOOCs, in many practical domains, whether involving repairing electric circuits, performing basic chemical reactions, or engaging in simple botanical, agricultural, or biological tasks.

Although we have stressed the importance of training in practical skills, there will inevitably be a need for and interest in scientific observation as well. It will not be possible to ensure that the first settlers will be equipped, talented, trained, and educated in all practical, technical, engineering, and scientific skills, not to mention psychological and psychosocial skills. But some training in observation, recording of such observation, and basic testing will be useful and possible. When Darwin set out on his first long voyage of discovery, he was not, formally speaking, a scientist. But he did go through some basic training at various scientific laboratories in the art of observing, analyzing, collecting, and preserving samples of what he encountered. This minimal training was sufficient to be of great use. Of course, we cannot all be Darwins and transform ourselves into famous scientists, but the first Mars settlers will certainly be trained in the necessary scientific basics of observation, sample collecting, and measurement. These basic skills should be of interest to the audience and those explicitly wishing to learn them in order to enrich their everyday lives. How much more interesting is it to go for a walk and notice what is really around us? A world of wonders can open up, which we never dreamed existed. We can all profit from such basics in observational skills. Our lives become richer and we are able to experience dimensions of the world around us we did not even know existed, or we did not dream of understanding.

Surely courseware in such valuable and practical training will be of interest to many who wish to learn such basic skills. This could also be part of MOOCs emerging as a result of preparing the new Mars settlers. Creating such MOOC courseware could become a joint initiative of organizations with expertise at making such courseware, educational institutions whose domain of teaching coincides with the subjects of training, and the Mars One team.

Mars One: Its Impact on Audience Members

Members of the audience who view broadcasts about the preparations for the Mars One mission will come from all strata and corners of society, whether children in their early formative years, young people still dreaming about what they would like to do as professionals, educators looking for material to inspire their students, politicians deciding whether or not to support a human space-exploration program, or taxpayers voting for those politicians to enter office.

Jules Verne's books inspired many innovators (and sometimes, the governments who funded them); some of the most famous enthusiasts who played a key role in the spaceflight revolution were very much inspired by him, such as Konstantin Tsiolkovsky (Russian), Hermann Oberth and Wernher von Braun (German), and Robert Goddard (American), all of whom became the pioneers of rocketry and human spaceflight. Later on, the launch of the first satellite, Sputnik 1; the first man in space, Yuri Gagarin; and the first man to traverse the surface of the moon, Neil Armstrong, inspired many young people to study hard, and not only excited almost the entire world population but also encouraged politicians to invest generously in human space exploration. This author hopes that the filming of these first steps in preparation toward human settlement on another planet will likewise inspire the peoples of the world, whether students or politicians, to support and invest in the human exploration of outer space, a collaborative effort of people from many nations demonstrating the advantages of working peacefully together on a common goal and dream, rather than investing in warring against one another or preparing for war.

Dr. James R. Kass *has been working in the field of human spaceflight for more than thirty years. Inspired by the first ventures of humans into space in the 1960s, he studied physics and mathematics, gaining a BSc, MS, and PhD (Canada, USA, and England). He was an investigator on both the first European Spacelab and the first German Spacelab missions in the early 1980s in the field of neurophysiology, training the first German and Dutch astronauts. He also trained cosmonauts flying on the Mir station and astronauts of the tragic STS-107 Shuttle mission.*

He has worked with cosmonauts of the former Salyut space station and astronauts of the first US space station, Skylab, and carried out research in several isolation experiments investigating psychology of long-duration isolation. He has taught in the fields of neurophysiology, biophysical research, and space psychology at universities in Germany, France, Spain, Canada, and England.

INSIDE THE MINDS OF THE MARS 100

This section is authored by the Mars One candidates. Before proceeding to Round 2 of the selection process, each was given a questionnaire to answer as part of the assessment of his or her candidacy. The answers to these questions opened a window into the hearts and minds of the candidates, revealing their dreams, personal reflections, and fears. This section brings together replies from the one hundred applicants who successfully moved on to Round 3.

Mars One applicants were asked about stress and fear, interactions with peoples from other cultures and with difficult personalities, and how being chosen for the mission would impact their families. Each question that the candidates answered is reproduced below, followed by a selection of their responses, which have been edited to preserve anonymity.

Through these excerpts, we hope to offer a glimpse into the lives of some of the brave women and men who are willing and eager to leave everyone and everything behind on Earth to embark on the risky one-way journey to Mars. A heartrending mosaic of stories grippingly told, they reveal lessons learned, gratitude for that which has already been given from friends, family, and from their cultural experiences, self-acceptance, foundations of the strength needed to cope with difficulties and challenges, and strong optimism about the future.

Following the candidate replies is additional information about these participants from a statistical perspective: the gender ratio, their ages, their nationalities, and their professions. As you'll see, these one hundred finalists reflect a wide swath of humanity—people, perhaps, who are more like you than you'd expect.

—The Editors

Question 1: Mission Mars would require you to be away from your family and friends forever. What impact would this have on you?

"Leaving behind my family and friends would be the hardest aspect of the mission. I don't know exactly how it would affect me; perhaps I am unable to imagine it vividly enough to make it feel real. I would be suspicious of anyone who was confident that they knew what their reaction would be this far in advance. The best answer I have is that it would be a huge sacrifice, but that huge sacrifices are called for when the prize is a whole world.

"I know that I would still be able to communicate with my family, even though we could never have a real-time conversation again. The emails and videos we would send back and forth would be like the rare letters that crossed the Atlantic centuries ago—messages from family members who did not expect to see one another again, but who could still share news and simply reassure each other that they were okay. So I suppose that I wouldn't think of myself as 'gone,' but rather as present in a different way.

"This question points to one of the most interesting paradoxes of the Mars One project: The first astronauts need to care intensely about our species, but be willing to leave most of humanity behind for the rest of their lives. They need to be highly socially adept but willing to limit their social interactions to a few people for many, many years. I believe that identifying the personality types that are best suited to these apparently contradictory requirements is one of the principal challenges facing Mars One."

—Mars One Applicant

"When we leave for Mars, I will have been married to my beloved wife for thirty-five years. She has been my heart and center for all my adult life. In 2022, my children will be in their late twenties, their lives just beginning to open into their full flower, perhaps ready to have children of their own. I can only imagine how hard leaving them behind will be.

"In the nineteenth century, when millions of people were emigrating from Ireland, although they might receive a rare letter from the Old World, they also faced the near-certainty of never seeing their friends and relatives again in the flesh. Around this time, they began celebrating what came to be known as an 'American Wake.' They would have a final party in which those leaving and those staying behind could share a final drink, have one last dance together, and say a final farewell. I think my loved ones and I will need the closure that only a 'Martian Wake' will provide. After that, email and laggy video calls with my family and friends will have to suffice to keep love strong.

"While it is a difficult thing to be sure of (and also a hard thing to admit), I think there are certain emotions I feel less keenly than most people, particularly anger and sadness. And although, previously, I've never had extended occasion or desire to test the theory out, perhaps this quirk of mine will help me deal with the separation from my wife and kids. I hope."

—Daniel Carey

"Friends may come and go, but family is not something that changes whether I am in the next room, away at school, or on Mars. There are a couple of precious friends that I count among my family and, like family, they are with me wherever I go.

"I really cannot say exactly what the impact such a separation would have on me as I have never before been to Mars. It certainly would not be easy, but this is my dream, and I know that my friends and family would understand this. I have about ten wonderful years left with them, whereas some people have much less. I count myself incredibly lucky to carry such amazing people with me in my heart."

—Mars One Applicant

"My line of work has taught me to live mostly in seclusion with little contact with my family. I would miss them to tell the truth, but I have adapted to being away from them for long periods of time. Yet I will live with them in my memory, because I know I can do this job."

—Mars One Applicant

"I know that going on such a mission would be hard on my family; I care much more about how they would cope than myself. I love my family; they have always supported me in doing what I have strived to accomplish in life. I know they would support me in this, too. Before leaving to go to Mars, I would ensure that my family was well provided for, and thank my friends for all the help they have given me through the years.

"Going on this mission is something that I feel I have to do; I have to do it for everyone back home, my family, my friends, and the entire human race. I want to give something back to them after all they have done for me—that is what true friends do for each other.

"Doubtlessly the first few years would be hard, but I would not be alone, I would be surrounded by some of the best that humanity has to offer. The dust storms may rage on outside, but humanity's best can weather any storm, because they rely on each other. They will be the first family on a new world."

—Mars One Applicant

"Absolutely I would miss my family and friends. Any sane person would long to reestablish connections of strong bonds severed by misfortune or distance. However, with that option absent on Mars, I would turn to what I have learned to do over the past several years. Since I left home for college, I have visited my family and friends from back home roughly once per year. Visiting more often is a financial impossibility. Thus, whenever I feel a desire to see them, I stop and reflect. I remind myself of where I am and who surrounds me. It then occurs to me that I have family with me wherever I am. My roommates of several years are as much a part of my family as my parents and sister. By the comfort and joy they provide, I trust they will help me persevere, and I know they know that I would do the same for them. So I imagine I am sitting on Mars and suddenly cannot resist longing for those back on Earth. Certainly, it would bring a great deal of pain knowing I could never see them again—to say otherwise would be foolish. Yet if I can now remind myself of the familiar bonds I share with my roommates forged over three years, I can only begin to imagine how wonderful it would be to support and be supported by my new family, with whom I will grow for eight years of training and beyond."

—Mars One Applicant

"I know being away from my friends and family may affect me after some time of being on Mars, but I have already considered all the repercussions of my trip and am mentally prepared for this epic adventure. My independent nature has led me to embark on my own adventures to other cities, and I have always coped well when being away from friends and family for extended periods of time.

"My adaptability coupled with my positivity are traits that I believe will serve me well when dealing with the challenges that may arise from living life on Mars. It's also important to remember that on Mars I won't be living a solitary life—I will have three fellow teammates living with me with whom I will eventually grow close enough to call family. I will forge a new relationship with my fellow three astronauts after the years of training together, and the four of us will forge a strong family-like bond that will help us overcome any obstacles."

—Divashen Govender

"This would be one of the greatest challenges. I communicate sporadically with my family, and love the catch-ups when we visit each other in person. I do a lot better with friends although, again, I can be undisciplined about keeping in touch because I know that eventually I will see them in person. I'm also quite tactile with people I'm close to, and require one or two very close platonic relationships for the sake of sharing my most personal thoughts and feelings. Coping with the Mars mission will require more discipline from me as to keeping in touch with friends and family on Earth, but as far as necessary intimacy goes, I imagine that I will become very close with my future crewmates. Like family to one another, they will be people I can trust, share secrets with, and have platonic physical contact with (I'm talking hugging, hands-on-shoulders sort of thing—I thrive on that). Communicating with friends and family on Earth will become an important part of my regular routine aboard the ship and on Mars, but it's something I'll be able to get into easily. I spent a long time growing up in remote locations, so communicating with friends over the internet is something that I'm familiar with. I also have several good friends online who I have not even met physically, so relationships with people purely over the internet are not new to me. My family is already working on ways of better keeping in touch when not together, and we're all enthusiastic users of social media, which really helps!"

—Mars One Applicant

"Supposing that the family is one of the most important things in my life, I (un)fortunately learned that everything changes, and sooner or later partings are inevitable. That said, I confess that if it will not be Mars, I'm going to move to Australia or the United States, so the decision to leave my family and friends was still present before I heard about Mars One. So there isn't much more to say; I love my family and my friends, but they know me, so I feel they would understand my decision; they would understand that travel into space and colonizing a new planet was my sincerest dream since I heard for the first time the word *stars*. I'll miss them, and they'll miss me, of course, but it's a sacrifice that I can make, because they brought me up for this; they were never opposed to the realization of my dreams."

—*Mars One Applicant*

"It would be terribly difficult to say good-bye to these people, and I would miss them tremendously. However, it is the nature of people to come in and out of our lives, regardless of how much we may love them. Distance—even all the miles between here and Mars—won't decrease that love, so it seems a small price to pay for the adventure of a lifetime.

"That being said, I also have every confidence that—if chosen—I will be in the company of some truly exceptional people who I may someday be able to count as my friends and family. Who knows? I choose to be optimistic about that point, rather than dwell on the costs for this fantastic adventure."

—*Mars One Applicant*

"I love my friends and family dearly and I know they love me, and that wouldn't change no matter how far away we were from each other, be it another country or another planet.

"Moreover, it's exactly because I know that they love me that I also know that they would rather have me far away and happy and useful than close by and miserable and not fulfilling my true potential. It's that old saying: If you love something, set it free.

"Therefore, whatever impact the separation would have on me, I just know that I would be able to use their love and support to strengthen my resolve."

—*Sara Director*

"My family and friends have had a huge impact on the person who I am today. My parents instilled within me the desire to learn, achieve, and explore, to find out what I want and pursue it indomitably. My friends taught me how to live with new and unfamiliar customs, how to be a better person to other people, and how to have fun.

"What I gained from them has helped me to deal with difficult changes. It has prepared me for tomorrow, and given me the strength to move forward even when times are tough.

"So, will it be hard on me to leave them?

"No, I do not think it will.

"I have been away from my family and friends for extended periods before with little to no contact. In some ways, going to Mars will give me more opportunities to keep in touch with them than I have had in the past, when I was out on the road. I will still be a part of their lives, and they will be a part of mine.

"I know that they would not want me to pass this up.

"When I told my parents I was applying to go to Mars, there was a long moment of silence before they said something to me I cannot forget. They said, 'You never cease to amaze us.'

"I will have that and everything else my family and friends gave to me, no matter how far away from them I am, and I will be able to use this knowledge to build a new life on a new world. I will not be leaving them behind forever—I will be taking them with me, close to my heart."

—*Mars One Applicant*

"I have been away from my family ever since high school, when I decided to go to another continent and pursue my studies. After my studies I decided to permanently settle on another continent [and] managed to adapt to this new country, which I now call home. I have an independent personality. I do not have a family of my own but my parents still live in Romania and, ever since I first came to Canada, we have been living oceans apart, only seeing each other once a year. This situation taught me and helped me understand that achievements in life require sacrifices. Friendship is important, and all my friends who know me well know that taking this opportunity of going to Mars suits my character perfectly. From my perspective, I want to be part of the team to attempt to set the foundations of humanity on a different planet. I think that would be my ultimate achievement in life, and I would like to be part of both the history and future of humanity."

—Andreea Radulescu

"I would be lying if I didn't say this would be the hardest aspect of the mission for me. I am very close to my family and not having the option to hold them would be very difficult.

"However, my family's boundless love over the years has given me the inner strength to stand on my own two feet.

"I will adapt on Mars and receiving video messages from my family will become my new norm."

—Mars One Applicant

"When talking to friends and loved ones about my dream of applying to the Mars mission, I received a wide range of reactions. Some think it is crazy or say it's selfish. Others call it a suicide mission. Some wonder why I would give up my life for this. And those who know me best often give a mixed reaction of sadness and understanding.

"When I told my daughter about the mission, her first reaction was, COOL! When I explained that there was a catch—that it was a one-way trip—she looked confused and disappointed. 'Oh,' she said as it sank in. After a moment, she looked up at me with a sparkle in her eye and exclaimed, 'But you'll get to go on an amazing adventure!' It was then that I realized just how much my little girl had grown up and really understood me.

"No matter how much I love my family and friends, especially my daughter and grandson, and would miss them all dearly, I am willing to accept and embrace this feeling for a chance to explore and colonize Mars. I would cherish the time I have with them over the next ten years before leaving Earth to cuddle my grandson, to say the things that need to be said, and to share their smiles and hear their laughter. And I would keep those memories close to my heart as I take the journey into the next phase of my life."

Question 2: Describe an incident that frightened you. What was the impact it had on your behavior?

"Being frightened is not just something that happens in response to big shocking events, such as fear for one's life; it's something that is an innate part of who we are. It's our mind's way of showing that we care!

"Of course I get frightened. Every time that I have to give a public talk I feel frightened—what if it went wrong? But I feel the important thing to do in such a situation is to acknowledge you are frightened, but not to let it influence your behavior. If I have an important exam, it's vital to not let fear over the result get in the way of a calm, logical approach to the situation. I take a deep breath, remember all the preparation that I have put in, remind myself of just how prepared I am and that the fear is unjustified, and then I allow myself to relax.

"It's important to reason with yourself and accept that being frightened is part of who you are, but not let it negatively impact your behavior. Fear can be overcome with even the briefest internal dialogue. I examine the factors that led to the fear, then plot out a rational course of action to mitigate its effects on my external actions."

—Mars One Applicant

"About two years ago, my grandfather, a very active man, had a sudden stroke and almost died. The idea that someone could be fine one minute and then be dead the next hit me hard, and it scared me. I started to think about my life and asked questions like, 'What have I accomplished with my life? If I died now? How would I be remembered?'

"The answers to these questions scared me just as much as the questions themselves. Because the answers were 'Nothing.' I would be remembered for nothing. Of course, they would say he was a smart guy and he was a good son and brother. But they say that about everyone. This was the moment I decided to live, to take chances—like applying to go on the craziest, most amazing, world-altering adventure called Mars One. The realization that I was just letting my life pass me by was the reason I finally moved twenty-two-thousand miles away to do a job I always wanted, and it's the reason I never settle for second best anymore."

—*Mars One Applicant*

"One night my troop camped in a wooded area two kilometers off an access road through thick, ankle-twisting mud. An hour after, a member of my section told me (as a senior recruit) that Joe was sick. When I found Joe, I immediately sent for help—it was near freezing and he'd stayed in wet clothes. His hypothermia was advanced enough to prevent him speaking or moving. Staff arrived and six of us ran back through the mud, carrying Joe in a tarp to an ambulance.

"All I remember is being terrified I'd trip, or drop the tarp from the pain of holding it one-handed. In either case, Joe would have died. The ten minutes it took to get to the ambulance were the scariest of my life—I was terrified my mistake or weakness would end my friend's life.

"We got Joe to the ambulance and he was stabilized. Later that night I'd save the lives of two others who'd not changed their clothes, working for hours to keep them awake as the hypothermia lulled them to sleep. It was a turning point for me. I learned that what truly scares me is the thought of letting down another in a moment of crisis."

—*Mars One Applicant*

"The recent bombing in Boston was only a few miles from where I live. I know the area well since I went to school in Boston and frequent the city often. I think about those who were affected—the families that lost their loved ones, those that are surviving with terrible injuries, those who experienced the explosions, the injuries, and the terror. I think about how scary it is to let my children out of the house, out of my sight. But the counterpoint of stifling them in fear of some potential unknown threat seems worse than living life knowing there are risks and some very bad people."

—*Mars One Applicant*

"I was supposed to go into the crawl space beneath my house to clean out some old debris that was blocking one of our pipes. However, looking down into the shadowed, dirty hole that was barely big enough to fit one person on their stomach, I could clearly see the numerous populated spiderwebs that I was about to get up close and personal with. I could imagine their occupants dropping down into the folds of my clothing, nestling between the strands of my hair, waiting for an opportune moment to bite while I crawled through their nests.

"As an optimistic person, I'd always tried to see the silver lining on each rain cloud.

"No matter how hard I tried, however, I could not find a silver lining to having spiders use me as their transportation. The possibility of spider bites was . . . not a reality I wanted to face.

"That was a first for me. Fear was not an emotion I was used to.

"However, labeling it as 'fear' only made me more determined to succeed. If I couldn't stop being scared, then I had to work around it and take one step at a time—putting one foot into the hole and then another, until I was under the house and reaching for debris. I focused on the task at hand.

"It taught me that while preparation can help us to face the things we fear, at other times planning is only the start. Sometimes, in order to finish, we have to be aware of the threat and forge through anyhow, doing what needs to be done even if it scares us."

—*Mars One Applicant*

"During my first three years of high school, up until the time my family moved to a new town, I was bullied relentlessly by three different popular groups of girls. Some were only catty by spreading rumors and generally making sure that nobody would accept me, but there were also a couple of girls who would do, and threaten to do, worse. I had my schoolbooks ripped up, was spat at in the face, and threatened physically via text. I was terrified to go to school every day but not brave enough to complain to my parents. Luckily, my second high school was completely different and I wasn't treated like that again. The impact of being bullied and frightened is that I avoid gossip at all costs and try to always act the opposite of how they did. If someone looks left out or lonely, I will actively try to include them in conversations and be genuinely friendly toward everyone."

—*Mars One Applicant*

"I was fourteen years of age when I watched as my mother died slowly of cancer. Her death frightened me. Yet the impact made me feel lost in my own world, up until I realized that now I had to take care of my father. It straightened me out, and I got working on making sure to be the strong one in the family, to work for my future, and make everything as easy as possible for my dad. In short, it made me stronger and it brought me closer to God."

—*Mars One Applicant*

"When I was about fifteen years old, my friend and I were on a bike and surrounded by six people, who cut our way off. It was on a small trail. On the left was the forest and on the right side was a steep scarp to the lake. Some of them were aggressive and my friend went silent behind me. So I had to protect my friend. I tried to talk to the aggressive ones. But they did not want to talk and hit me instead with a bottle on my right temple. I blacked out on impact and woke up on the ground.

"Four of them had escaped. The remaining two were shocked and tried to apologize. I yelled at them and was agitated that nobody had come to our aid. They just wanted to have their event—hurt somebody—and we were there. I realized there does not have to be any reason to be targeted."

—Robert P. Schöder

"The day I landed in India, no one came to pick me up at the airport.

"All I had was the phone number for the person who was supposed to pick me up. So I activated the global service on my phone and called the number. On the other end, someone picked up, but they didn't say anything. In the background, I heard crying.

"I hung up since no once was answering when I repeated 'Hello?' over and over. I had printed out all of the correspondence and scoured it to find another number. A second number got me in touch with Mukesh, who came to get me. When Mukesh arrived, he told me that the woman who was supposed to pick me up had died that morning, hours before I landed.

"While the situation was frightening, my proximity to death frightened me even more. But it taught me that anything can happen to uproot and change what might be expected. This experience taught me to go along with circumstances and, instead of worrying, to prepare and act when presented with an unexpected challenge or misfortune."

—Mars One Applicant

"When I'm not climbing on cliffs, I attend a rock climbing gym in the city. There are pulley systems at this gym that allow a person to clip their harness onto it and climb up a high wall and repel down without assistance. One day I stepped up to the wall to see how I would start a certain climb and I forgot to clip in. Next thing I knew, I was at the top and realized there was no rope attached to me. I was frightened, but I took some deep breaths and began to climb down. Climbing down is always harder. This particular climb was difficult and near the end of a two-hour session, I was already tired. I made the climb down safely, but the incident has definitely stayed in my mind. Now I always check multiple times to see if all of my equipment is in place before I approach the wall. I also now climb down every time, just for the practice."

—*Mars One Applicant*

"After entering the United States, I joined a computer science MS program, which was new for me as I held a BS degree in electronics. In the first semester, I got *C* grade in one of the core courses, which endangered my assistantship. Without the assistantship, it was difficult to survive for me because I did not have any other financial support. Fear of going back to India without completing my education frightened me the most, as it was my most rebellious decision against my family. Although it never caused any visible or radical change in my behavior, I did lots of hard work to overcome this fear. I took the course again, and started studying the whole night in the laboratory and coming back in the morning after few hours of sleep. I studied this course from all available online education programs and video feeds. I reduced my extracurricular activities and started more and more meditation so that I could concentrate on the subject properly. Finally, I aced the course with the highest grade."

—*Mars One Applicant*

"During a reunion with my college friends, four of us were out in the early morning swimming in a pool beneath a waterfall. One of them, quite drunk at the time, decided to start climbing. We assumed he would go no farther than the first ledge, but he quickly climbed to a point where he was out of our sight and earshot (due to the noise of the falls). The treacherously slick rocks terrified me (and had discouraged me from climbing them the previous day), but my concern for my friend's well-being was the overriding factor. Therefore, despite that fear, I cautiously picked my way up to that first ledge until I caught sight of him. From that vantage point I was able to shout to him, and managed to talk him down from trying to climb the entire waterfall. I didn't realize how afraid I was until after he was safe and I had to climb back down myself, but after carefully selecting a different path down, we both made it back to the bottom uninjured."

—*Mars One Applicant*

"A year ago, when I was walking, I got lost in a military-restricted area. I only had water and did not know where I should go. After three days, I found the way out. For me, the effect is always the same—the road goes forward. Get up and continue to go. Giving up does not exist."

—*Mars One Applicant*

"I was about fifteen years old and I loved going to the neighborhood swimming pool even though I was a novice swimmer.

"This particular day, I was alone and decided to test my not-so-good swimming skills because there wasn't anyone around that would make fun of me.

"I ventured toward the deep end, and suddenly lost control. I started drowning. That's the last I can remember. Apparently, I was saved by a lifeguard that happened to walk past. I was back in the pool the next week, and today, I am a real good swimmer, as I did not let that incident deter me from learning how to swim."

—*Ighodalo Eromosele*

"I was recently asked to give an interview to the BBC about my views on the Mars One mission. I had never given any kind of press interview in my life and I was frightened. What if I came across badly? What if people laughed at me?

"I reasoned that giving this interview was important, not only to help promote the Mars One mission but also to allow me to grow as a person. If something is important, I have never backed down from it, and so I gave the interview.

"Mars One is allowing me to break down barriers even now. I hope there are many more to smash."

—Mars One Applicant

"Military tanks passed near the house, and it was very frightening to hear the huge tanks and the sounds of impending military action.

"My first thoughts were of my children and their future. I contacted friends living abroad to try and find out how to flee the country with my children, in case of danger. In addition, my ex-husband and I supported a group of demonstrators who were attempting to barricade buildings. They stayed overnight in the streets, and we brought them food and warm clothes."

—Mars One Applicant

"Having taken a risk to take the road less traveled, I found myself homeless during the winter in Canada with no money and no job. Needing a place to stay, I registered with the Salvation Army, and nighttime provided some of the most frightening experiences for me. All the men slept in one dorm with single-sized beds, and belongings were kept underneath the beds, mostly in plastic bags. During the day, the people I encountered were tough, dangerous—you wouldn't want to cross them during the day, let alone in a dark alley at night. Yet while they slept, their guards were down and even the most hardened thug had nightmares. I used to sleep with headphones on because I was afraid of the noises they made at night, and of the things they said in their sleep. (As a night owl, I'm usually awake at night when others are sleeping.) It scared me. But when my batteries died, I had no choice but to listen to the noise and the words. And that's when I heard some of them crying. These guys who would cut you if you looked at them the wrong way cried in their sleep. Even after I got new batteries, I stopped listening to my headphones at night while I was at the shelter. Not only was it a way to face my fear, but it made me more sensitive to them as human beings, and I treated them with more dignity and respect from that moment forward. They never knew that I heard some of the frightening things they said in their sleep and they will never know the impact they had on me becoming a better person."

—*Mars One Applicant*

"I was flying a plane with very strong crosswind. I recall that I had a sudden loss of lift due to wing tip. I realized it was a bad situation to be in and took appropriate action by giving full throttle and countering the wind gust."

—*Mars One Applicant*

"I was hiking alone along a wilderness trail in the mountains when I began to get severe leg cramps. I was out of water and I knew it was due to dehydration and electrolyte imbalance. But I also knew I had to hike uphill for another mile or two to reach a spring. I pressed on, enduring leg cramps with an intensity of pain that I had never experienced before. At one rest stop, I leaned against the cliff face and slipped down on to my back. I found that I could not move my legs at all, as though they were paralyzed.

"Being incapacitated, with no one around to help, had never happened to me before. I started to fear what might happen to me. I had no choice but to lie there and think about my situation: The sun was going down so there would probably not be another hiker for hours. I needed water but water was about one mile up a steep switchback mountain trail. I tried to roll over but didn't have enough energy at first and just rocked back and forth on top of my pack. I got the image of me being a helpless turtle rocking on its back, flailing its little arms, and I just started laughing. Laughter and rest gave me a clearer head and enough energy to sit up and massage my legs. Eventually, I painfully reached the spring in the dark and topped off my water bottles. In times of distress, my wry sense of humor helps me to put the problem in perspective and my mind, once relaxed, starts to find solutions, or at least determines to cling to hope."

Question 3: Describe an event that increased your stress level dramatically. What was its impact on your behavior and what did you do to regulate it?

"When I was sixteen, my stepdad left my family while my sister, my mother, and I were out at the cinema. We returned to find the house stripped bare, everything was gone, even the telephone line. It felt like my entire world had turned upside down. This alone would have been shocking and stressful enough, but in the previous year my grandma had died and my dad had run away to New Zealand. You can't change the past, but what you can do is save what you have, protect those you care about, and deal with what you've been dealt.

"I had to remain strong for my little sister, who was struggling to cope with all that had happened, while at the same time working toward exams that would determine my future. The trick to dealing with a stressful situation is to acknowledge reality, identify the factors that are causing the stress, then either acclimatize yourself or remove the sources of stress.

"The years that followed were hard, but I supported my family through them and, equally, I knew I could rely on them. Together as a group we dealt with the stress.

"In the end, I managed to emerge successful despite all the odds thrown against me. I was the son of a single-parent minimum-wage family, living in poverty in a first-world country, and yet I managed to get into one of the world's top universities to study the subject I love. I've found that the real key to coping is believing—believing that you can overcome any obstacle."

—Mars One Applicant

"I survived my house having its roof ripped off by a tornado. I survived crashing on a major highway in a blizzard for days, stranded hundreds of miles from home. I survived my final college semester, finishing my research project on biosequence similarity searches while preparing for a cross-country road trip with a half-done pair of full-sized mechanical wings.

"Disasters, heavy workloads, and speaking onstage to thousands of people have not fazed me.

"I don't faze easily.

"Auditioning in front of a small group of people for the high school play, however, used to come close.

"I'd fret about it for nights, wanting so badly to not let down my family, friends, or myself. I worked hard to practice my lines and my singing, knowing that when I got nervous it could be hard for me to transition from my chest voice to my head voice. I prepared until I was so comfortable with the material that I would even recite it in public with my friends.

"When I stood onstage for the audition, my hands were freezing, and I was shaking, but I did not back down, singing as hard as I could despite how fast my heart was beating.

"It was always worth it.

"Getting to be a part of each play was a chance to make something bigger of myself. It didn't matter that it was stressful to get in, because the hardship meant something. It taught me how to practice, how to persevere, and how to follow through with pursuing my dreams no matter what the odds might be.

"I wouldn't be applying today without that."

—Mars One Applicant

"When I was deployed, I was placed in command of sixteen airmen and seniors enlisted in Kuwait and the Iraqi border. We had a full plate with three missions, and I dealt with the stress of prolonged working hours, the expectation of command, and the unknown dangers of working in the desert by exercising and reflecting on how my stress could only be determined by things I could control. I exercised, reflected, and slept, and the stress went away."

—Mars One Applicant

"Well, this one happened during my first pilgrimage fifteen years ago. A lot of things happened during that trip, too much to put them down here. The combination between high physical demand, inadequate meals, and some personal feelings between the members in the group I was traveling with put me in a state of high emotional depression, which almost led me to commit suicide.

"That experience taught me a lot about myself and mankind and about the world, and it fully changed my point of view in many aspects of life. I learned where my own limits were and how to avoid them; I learned about patience and about love and friendship. I learned about how to handle my own feelings and to recognize the difference between the act each one requires, and if it will be really good or not for me and the ones I like or love, and I learned the difference between self-esteem and egotism. That experience taught me also about how other people feel—how they can be hurt or healed—and how to immunize myself against their bad moods (if they have them) and live happier.

"Finally, this experience changed my way to see life in general. I changed almost all my philosophies about life in the months after. I think they are pretty solid actually, as I have tested them several times and they have never failed me."

—Mars One Applicant

"Recently, I have experienced several life-changing events that have caused stress in my life.

"One stressful experience that stands out is when my stepfather became critically ill. Because of conflicting reports, the doctors could not, or would not, tell us what exactly was wrong with him. After almost a year of tests, doctors' visits, and trips in and out of the hospital, we watched helplessly as he lost his life to lung cancer.

"Sam, as we lovingly knew him, was a strong and vibrant man, a proud Navy diver who loved to play pranks and had a contagious smile. It was so difficult to watch this man wither away, and to endure such pain, and to not be able to do anything about it.

"Caring for a seriously ill loved one, without knowing what was wrong, and not being able to do anything about it caused so much stress for our family. Finding balance between trips to the hospital, being strong for my stepdad, supporting my mother during this difficult time, helping with the final arrangements, trying to meet the rest of Sam's family's wishes—all while raising a teenager, along with other work, school, and volunteer commitments—felt like a monumental feat.

"At times, the stress felt like too much to bear, but I had to persevere, to be strong, and have faith that we would get through it together. Reflecting on the positives—the fond memories, knowing that Sam was comfortable at the end, and that he was finally at peace—made it easier. Finding inner strength helped me find peace."

—*Mars One Applicant*

"Once, while serving as a resident assistant at a university, I had to negotiate a hostage situation. Despite all my efforts to get outside help, I was left in the situation alone with the aggressor (who was armed and dangerous) and the hostages. It forced me to listen, not only to the aggressor but also to myself. I had to trust my instincts in the situation and I had to be sensitive to the needs of the hostages, namely their safety. After an hour of negotiations, I was able to diffuse the situation without harm to anyone, and this was accomplished only by listening to myself and doing the opposite of what was expected. The only thing I could tell myself during that time was to do my best and let the universe take care of the rest. I had to realize that it was possible that people could be seriously or fatally harmed. I had to accept my limitations and work with what I had available at that moment. I had to remain calm for the sake of those around me."

—Mars One Applicant

"I traveled to a North African country when I was in college. It was my first time experiencing an Islamic country. On the street, every three meters, I was sexually harassed by local men.

"I felt much stress and did not go out to the street. However, I remembered that I came to see the country and enjoy myself. To solve this, I went to a local shop to buy a scarf to cover my head and local traditional cloth to wear to mix with local women. The next day, I went out on street again but with local attire with a scarf to cover. I did not get bothered by local men. I could freely walk at local market and everywhere, but I got a new problem: I was offered two hundred camels for marriage."

—Mars One Applicant

"One occasion that comes to mind when I think of being the most stressed is during a flight lesson under strong gust conditions. During my lesson, I had trouble performing basic maneuvers and my instructor was less than pleased with me. The gust was causing unexpected turbulence on my Piper Cherokee single-engine plane, and my instructor was yelling at me for not being prepared. My stress level increased rapidly, my head was throbbing, and I began to read erroneous altitudes and was on the wrong heading. I understand how dangerous it is to lose your sanity in my circumstances. I asked my instructor to stop talking and I proceeded to land the airplane immediately. I took a few seconds to clear my mind and take a few deep breaths while I directed the airplane toward the airport for the descent. Right before contacting the control tower, I flipped my checklist to where I had previously written backup communication commands. Having my notes in order and being able to successfully communicate with the tower increased my confidence and brought me back to a state of control. As my confidence increased, I was able to have a successful landing despite the gusty winds on final approach. After the landing I spoke to my instructor about the way he treated me up in the air. He apologized and praised me for deciding to land the plane before our scheduled time."

—Mars One Applicant

"At age five my parents filed for divorce. At that age it was stressful, but growing up with my parents separated was much more stressful. I became more withdrawn and sometimes questioned if I was to blame. As time went by, I overcame this by focusing on all the things I had/have in my life as opposed to all the things that I didn't/don't. This has made me a more positive person and I always see the brighter side of situations. Becoming more positive gave my confidence a boost and this, in addition to my parents' reinforcement, made me realize I was not to blame and that the separation was a mutual agreement between them for their own personal reasons. My parents' separation also forged a sense of independence in me that I am very proud of. I am now a very optimistic person that has much confidence in myself and what I can achieve. Whenever I do get sad or down, it doesn't take me long to assess all the positives in my life and pick myself back up. As a result, I believe that you can achieve your goals, whatever they may be, as long as you push yourself to the best of your abilities and never give up."

—Divashen Govender

"Last year was likely the most stressful time of my life—I had gotten laid off unexpectedly from a job I loved, and I was unemployed for a long time. The stress did affect me tremendously, but I dealt with it.

"The most frustrating thing about that situation was a feeling of not being in control of my fate, since I was at the mercy of potential employers who may or may not have wanted to hire me. My solution to that was to focus on the things I could control: my résumé, managing my professional network and contacts, practicing for interviews, and essentially all of the things that focused on improving myself, rather than anything external. This worked well. I kept myself busy to distract myself from the stress, and doing so actually contributed to the solution. I got a great job that I truly enjoy, and I feel as though I've gained some valuable confidence along the way."

—*Sara Director*

Question 4: Describe a situation where you experienced interaction with a culture other than your own. What challenges, if any, did you face and why?

"The first time I visited a country other than my own was certainly an interesting experience. Though for most people such an experience would happen when they were young, possibly going on holiday with their families, I went alone when I was sixteen. I had been selected as one of three people to represent my country at the European Space Camp in Norway; it was my first real taste of the plethora of different cultures that exist in our world. But the thing that really struck me was how similar the people from these entirely different upbringings were. We shared common interests, common dreams, and common humor! I do have to wonder: In this globalized world of ours, are we are really that different from anyone anymore?

"Yes, cultures may be different, but people, individuals, now there is something that you can always relate to. In the past, nations in conflict would always try to demonize the opposing side because they knew, deep down, that it would be difficult to harm the other side because they were people just like themselves. You could put two people from different sides in a war together in a small room with no knowledge of the origins of the other and, despite everything, they could get along; they could work together. This is because they are human, just like everyone else. There is no such thing as different cultures when you really look at it, there are only people, only humanity—that is what I have learned."

—Mars One Applicant

"For a semester in college I studied at a university in Scotland. Upon arrival my entire world changed. The housing was different, as was the cuisine, accents, left-sided drivers, and overall social life. Even the lingo was not akin to my usual vocabulary; only a few laughs were required for me to realize *pants* actually meant underwear. I found it difficult to adjust to the fried fish and potato-saturated diet, not because I dislike potatoes, but rather because my digestive system had not braced for impact. But the biggest challenge I faced was adapting to European social life. Not only did I befriend Scottish students but also those from around Europe and, on the whole, they formatted their social gatherings around very late nights and alcohol, no matter the day of the week. My typical week focused on completion of work and schoolwork first and foremost, with the weekends being open only if all tasks had been finished. In contrast, their week consisted of partying hard until only enough time remained to complete assignments. And with no grade mattering but the final exam, it felt as if even the university advocated 'party first, work later.' As someone who partied very little, I nonetheless adapted to the new environment and, although I'm back to my regular self in the United States, I have to admit I had nothing short of a wonderful experience with those I met in Scotland."

—Mars One Applicant

"I visited Japan in 2009 with two friends of mine. It was something we'd always wanted to do, and we had a fascination with the culture, so we went in with an idea of the Japanese culture and mind-set. The biggest challenge, however, was communication. Very few people in Japan speak English well, if at all, and our Japanese was incredibly basic. A lot of communication had to be improvised with hand gestures when necessary, although Japan is very accommodating to non-Japanese-speaking people, with things such as picture menus commonplace.

"There was another level of abstraction caused by common hand gestures in Japan being slightly different from in the United Kingdom, but we made sure to pay attention and see the humor in the situation, which often allowed us to connect with the people we were speaking to in a positive way, regardless of not completely being able to understand one another. Our greatest strength, however, was working together. One of us had the greatest understanding of the culture and language, one had a lot of practical common sense and grounded thoughtfulness, and I had the confidence to take advice from the others and approach strangers with questions and queries, and the ability to quickly pick up how things worked (e.g., the transport system and how to order food in certain kinds of restaurants). Finally, we made some friends while there, who took us to places other than the usual tourist spots so we could get a better idea of everyday life in Japan."

—*Mars One Applicant*

"Although I stayed in a first-world country for half a month, the greatest culture shock I experienced was when I entered college. I grew up in a small provincial town in Southeast Asia, but my parents decided I would enter college in the big city, and they chose an exclusive Catholic school for girls. The first day came and pure shock overwhelmed me. Some students looked and acted so masculine that I never thought I'd find a girl hot. Girl-on-girl sexual relationships were real. Most had boyfriends, snogging was normal, and breaking your virginity didn't automatically make you a whore. High heels and makeup were part of the daily getup. And laptops weren't just for business executives.

"I grew up in a conservative environment, and all that was contrary to the ideas, lifestyle, morals, and principles I lived by and garnered in the earlier years of my life. On top of that, urban dwellers often discriminate against people with provincial accents, always assuming they come from blue-collar workers. Despite the immense differences and the slight discrimination, I never became a lone wolf. The best thing about new worlds is that you discover new things and get a clearer picture of the world. It was difficult at first—a teenager trying to fit in—but as long as you walk among people of another culture with the intention of understanding and comprehending, instead of outgrowing or opposing, friends will come to you."

—Mars One Applicant

"The biggest challenge was discovering how differently people treat each other. Many of the ways in which I was used to interacting with people simply weren't applicable, and I had to adjust my sense of social norms.

"Shifting my perspective like that seemed a bit unnatural at the time, but having that experience to draw on has actually enriched my interactions with people to this day."

—Sara Director

"I stayed with a Spanish family. I was paired up with my friend, and we were placed with a relatively old couple in a flat in an apartment block. Upon arrival, we were shown to our room, and it became immediately obvious that there was a significant language barrier. The couple could not speak any English, and my Spanish at the time was very limited. So, using smiles, body language, and our best attempt at the local tongue, we thanked them and settled in.

"The couple was very welcoming and acted as parental figures during our stay. They gave us keys, so we could come and go as we pleased, but made it clear we would need to be back for dinner.

"At our first meal with them, they brought some beautiful home-prepared fish to the dining table, to the horror of my friend, who was a vegetarian. [Using a] dictionary, we politely explained the situation but gratefully accepted the salad and potato accompaniment. Although it was awkward, it's amazing how the language barrier could be overcome with gestures and awareness of another person's body language and tone of voice. We showed them our gratitude and respect by clearing up after the meal.

"This trip was really eye-opening and memorable, especially when we left, exchanging gifts, and were told we were welcome back again."

—*Mars One Applicant*

"Too many people live in a cultural bubble and think that everyone thinks and lives like they do. They could not be more wrong. My most difficult cultural interaction was when I was assisting a couple from the Middle East. When I met them, I addressed both of them. The man quickly spoke up and explained that, in their particular culture, women are not to speak to males outside of their family, and have a male representative speak for them. I was raised that all humans are equal and I found it very hard to not direct my questions and comments directly to her. While I did not believe that their custom was right, I respected it and spoke only to him. Just as I would expect that man to respect my culture and treat my mother as an equal, as that is the way of my culture. Throughout my years of working with international clients, I have learned that the key is to understand that you don't have to 'understand' why a culture does the things they do. Most of the time their way of thinking is so foreign that you never will. What you must do is respect their cultural decisions and heritage, and accept them for who and what they are."

—*Mars One Applicant*

"I spent two years living and breathing the culture of Malawi, a small African country, as a Peace Corps (PC) volunteer. I was virtually adopted as the eleventh child by my host family. I lived in a mud hut on the family compound, fetched water with my sisters, and kept my own garden. My work helping the local milk-bulking group to become self-sufficient was done in Chichewa. Only a few of the members spoke English, which was useful at the beginning when I was learning the language. Even as I became a part of the local community and adapted to the culture—a key facet of a successful PC experience—I found there were things within the culture that I loved and things I could not accept. The culture challenged everything I considered normal, acceptable, and right. I learned to be more objective and to acknowledge my own cultural limitations; this was an internally difficult process but one that benefited me later. By living and becoming a part of Malawian culture, I shed many pieces of American culture that I had unconsciously adopted. In doing so I became separate, in a small way, from both cultures. It was not until I returned to America and then went to France that it struck me how different it was to look at the world and seek out what makes each culture different: There are whys behind the cultures that do not make sense on their own. In removing some of my cultural blinders, I feel that I have a wider view of the world."

—Mars One Applicant

"In my travels around the world, I spent lots of time in India. The concept of personal space is different there; the tendency is to get very close to another person. So close that, to the Westerner, it feels like an invasion of personal space. I had several ways to 'defend' against this invasion. My first line of defense was to pretend that I saw something interesting a few feet away. If that didn't work, I would excuse myself by saying that I didn't feel so well and that they better stay away from me unless they wanted to catch a bug. If I got to know them better, I would try to explain the different concepts of personal space.

"In general, it was important not to hurt their feelings and, at the same time, not suffer the invasion of personal space."

—Mars One Applicant

"Working on a big experimental setup along with some foreign colleagues, we had to make use of gestures as much as possible because of significant distance between us. It took us quite some time to realize that a simple gesture such as shaking the head can mean different things in some cultures. Now I spend more time getting to know the cultures with the representatives I have to work with side by side, especially if it requires time-sensitive actions."

—Mars One Applicant

"When I was in elementary school, some girls in my class made me feel badly by pointing out that I was 'different' from them. This was true; I came from a different ethnic background than they did. It was hurtful and confusing because I thought we were friends. I didn't see myself as different from them. After all, we were all human beings in the same class. I learned some valuable lessons from this experience: First, that it can be hurtful being cast as an outsider, and that I never want to make anyone feel the way I did; and second, that we should not be blind to our differences like I had been, and rather, accept and embrace them. I believe that as long as we behave with openness and a willingness to learn, we can find beauty in the differences and similarities among us."

—Mars One Applicant

"When a friend and I traveled to Tanzania to climb Mount Kilimanjaro, we met people that lived in a very different way than we were accustomed to. There were so many who lacked what we took for granted. Part of my trip preparations included studying simple, polite phrases in Swahili and taking pins and other souvenirs from the United States to share with them. The people that we met, not just the tour guides but also the people on the street, in the shops and restaurants, were very friendly. It was not hard to realize that people are people everywhere; they enjoy laughing and having fun and meeting new people as much as we do. A few that we got to know invited us back to their homes to share their hospitality and spend some more time talking with us. I had the opportunity to travel to many places around the world, and I have found that the exchanges with the peoples and cultures were the most rewarding and memorable parts of those trips."

Question 5: What type of personality would you have difficulty living with?

"I would have difficulty living with a person that couldn't accept help from anyone. When a group of people has different specialties, interests, and beliefs, you have to accept sometimes that other people are better than you at certain things, and that you can be enriched by positively engaging with them. It's especially important in a small community for people to work together, playing off their own strengths as well as the strengths of others. Of course, people have to be capable of acting independently, but it would be dangerous if someone split off from the group, particularly psychologically. I enjoy helping others when I can, but I equally enjoy learning new things from others; there has to be an open exchange and dialogue of ideas, and I would hope that the first settlers on Mars will appreciate that."

—Mars One Applicant

"I'd have a harder time living with someone who was rigid and unable to see new and inventive ways of perceiving the world around us. Oftentimes a person who is more invested in being right about their own viewpoint is unable to access the greater beauties of opening up to an unlimited vista of perceptional varieties—I think this is an asset to have in unexplored landscapes and I've been able to hone it through my own explorations. I did not realize life could exist in myriad ways until I began traveling—that not everybody has (or needs) toilet paper and that hot showers are actually a luxury, but never a necessity. Someone who would not be able to have a peaceful state of mind without the material excesses we're accustomed to would also be hard to live with, as their focus would be on the negativity of loss versus the optimism of discovery—a trait required to live on a planet of little material resources."

—Mars One Applicant

"Quite frankly, I'd never want to be with a person whose life revolves around him- or herself. I have never been able to put up with conceited people and people who think that they deserve all the sympathy in the world. It is high time that such people take a look around and learn from the lives of the underprivileged."

—Mars One Applicant

"I like confidence, but I don't like an inflated ego, and there is a difference. I find selfishness annoying at the best of times, but in a situation in which we need to trust each other, we need to be looking out for one another. Not to mention that we will be representing the human race, and this whole mission is about something bigger than just us. I would find self-obsessed people frustrating. Additionally, I am ground down by people who constantly complain about small things, and find their company quite stressful. Having a gripe every now and again is fine, but life on Mars will be hard work, and any applicants will need to accept that uncomplainingly, because the opportunity is worth the difficulties. I am very introverted, and struggle when people do not respect personal boundaries if I need some time with my own thoughts. Similarly, I can be uncomfortable around people who feel the need to talk all of the time. People who know each other well should be comfortable with each other in silence, and not feel the need to interact constantly. Finally, while I have no objection to other religions or atheism, I am a Christian, and while I am open-minded and have developed a fairly thick skin against mockery, I can't stand it when people assume that I am stupid or look down on me for being religious, because they do not appreciate the amount of thought and consideration I give on a daily basis to what I believe and how it fits in with the rest of the world."

—Mars One Applicant

"By far, persons with an explosive temper and those with deep materialistic pride never really mix positively with my groove. Hotheads are usually reckless and uncooperative. Most of the time, they will not accept logical explanations, suggestions, and advice from others. Whatever action they deem correct, they will force. I am a person who values the ideas of the people I work with. Seeing something from someone else's perspective gives me a better view, thus giving me a firmer grasp of the situation. And I like to take my time in discussing with my teammates the best and most efficient course of action toward a mission or a problem. For the aforementioned second type of personality, I just don't want to hear how much money or how costly their things are. To me, it's pointless to live a life dictated by brands and capitalism. In my opinion, these things keep people from things that really matter, such as spiritual growth, family, adventure, nature, and the list goes on. As to how I treat these characters, I don't necessarily oppose them, but I do my best to avoid their presence. Should interaction be required, I call upon my entire willpower to understand and [empathize] with these people and be as friendly as possible."

—*Mars One Applicant*

"The only type of person that I can honestly picture myself having trouble living with is someone who is consistently negative, or worse, not a team player. Personally, I'm a glass-half-full kind of girl. I have an endless supply of optimism and I approach challenges and obstacles with humor. I never call something a problem. Instead I put my nose to the grindstone and figure out how I can best contribute to a solution."

—*Kellie Gerardi*

"People who feel unsatisfied over everything, are hypercritical, never find happiness in small things, always one-sided, pessimistic, living on credits, uncommitted, faultfinders, suffering from low self-esteem, hyperexaggerate, and manipulative or dishonest. These people have no appreciation for your thoughts, creativity, or ideas and are always filled with negative thoughts. I avoid individuals who engage in regular and frequent gossip, give more importance to destruction rather than being constructive, and are biased toward someone instead of bringing everyone together."

—*Mars One Applicant*

"I don't think there is a specific personality that I would not be able to get along with at all. I think every person has a diverse mixture of personality traits, so there is always something that I can find to relate to with a person. If there is a will there is a way. I am highly adaptable and my humorous nature does help me relate to others. I think if a person just takes time to get to know another person, there is always a way to coexist. Most people do not seem to know their own limits and who they are, but if they are open enough to others, they can also discover themselves. Traits that might not agree with my personality are a lack of curiosity, no sense of adventure, no ability to think and act in stressful and extreme situations, an unwillingness to cooperate, and dishonesty."

—*Andreea Radulescu*

"I would find it difficult to live with people who lack compassion and empathy. I would also find it exceedingly difficult to live with a Hamlet—someone who cannot move forward because they are mired in their own thoughts and indecisions."

—*Mars One Applicant*

"I get along with people fairly easily. However, being a positive person myself, I can find 'negative' people challenging at times. I understand and value the occasional need to be critical or point out the improbability of things to prevent wasting valuable time and resources. I also understand that people, including myself, have bad days and our spirits can be dampened. I usually use my humor to lighten the mood or try to elicit constructive criticism to develop solutions where there only seemed to be problems. But I would find it most difficult living with a chronically dour person that spends most of their energy finding fault, placing blame, and avoiding attempts by others to help them see life's brighter side. On Earth, we have the luxury of avoiding people we really don't like, even getting a divorce from our most intimate partner, but in a small community—the only community—on planet Mars, it will be crucial that we can all be respectful of each other's personality types and learn to negotiate our differences to achieve the goal we all have in mind: a fulfilling life on a new planet."

Candidate Statistics

The candidates listed on their applications the following occupations:

Publisher
Stand-up comedian
Structural engineer
Project risk specialist
Teacher
Dental surgeon
Kitchen hand
Public servant
Anthropology student
Physics PhD student
Exploration geologist
Office manager
TV producer
Account manager
Technical staff
Apparel industry staff
Seamstress
Mechatronics MS student
CTO
IT administrator
Sports marketing specialist
Solar-monitoring software technician
Physicist
Graphics artist
General practitioner
Physics MS student
Astronomy PhD student
Science laboratory technician
Systems-integration manager
Astrophysicist
Archaeologist
Logistics manager
Space sciences student
Research and development engineer
Software developer
Biologist
Architect
Census-collector manager
MD student
Chef
Architectural project manager
Web administrator
Cashier
Writer
CEO international metal source
Emergency management officer
Marketing development specialist
Economics graduate
Geologist
Polymer chemistry PhD student
Particle physicist
Fitness trainer
Journalist
Geographer
Dance teacher
Space engineer
Store manager
Editor

Reservist
Ballet choreographer
Technical support
Scientific language editor
Quantum mechanics research assistant
High-tech executive
Actor
Aerospace engineer
Data architect
Beekeeper
Mars science-laboratory robotic-mission scientist
Radiation health physicist
Health administrator
Satellite start-up manager
Project coordinator
Senior administrative assistant
Civil engineer
Flood protection
Projectionist
Military university student
Outreach manager
Emergency medicine physician
Production geologist
Designer
Data analyst
Human factors and systems engineer
Garment design
Paramedic medicine PhD student
Political consultant
Education start-up founder
Law graduate
Physics graduate
Assistant manager
Quantum biologist

Candidates by Gender Identity	Total	Male	Female
	100	50	50

Candidates by Age	Total	Male	Female
Less than 26	16	7	9
Between 26 and 35	48	24	24
Between 36 and 45	23	13	10
Between 46 and 55	11	5	6
Greater than 55	2	1	1

Min. age: 20

Max. age: 61

Candidates by Highest Degree	Total	Male	Female
No degree	21	14	7
Associates	2	1	1
Bachelors	37	18	19
Law degree	1	0	1
Masters	29	12	17
Medical degree	4	1	3
PhD	6	4	2

Candidate Nationality by Continent	Total	Male	Female
Africa	7	5	2
North America	39	17	22
South America	3	1	2
Asia	16	7	9
Europe	28	17	11
Oceania	7	3	4

Candidate Nationality by Subregion (Nationality)	Total	Male	Female
Northern Africa	1	1	0
Southern Africa	5	3	2
Western Africa	1	1	0
North America	39	17	22
South America	3	1	2
Eastern Asia	4	1	3
Southeastern Asia	3	1	2
Southern Asia	8	5	3
Western Asia	1	0	1
Eastern Europe	9	2	7
Northern Europe	7	4	3
Southern Europe	6	5	1
Western Europe	6	6	0
Australia and New Zealand	7	3	4

Candidate Nationality	Total	Male	Female
Austria	1	1	0
Australia	7	3	4
Bolivia (Plurinational State of)	1	0	1
Brazil	1	0	1
Canada	4	1	3
Switzerland	1	1	0
China	2	1	1
Czech Republic	1	0	1
Germany	2	2	0
Denmark	1	1	0
Egypt	1	1	0
Spain	2	2	0
France	1	1	0
United Kingdom of Great Britain and Northern Ireland	4	1	3
Georgia	1	0	1
Croatia	1	1	0
Ireland	1	1	0
India	4	2	2
Iran (Islamic Republic of)	3	2	1
Italy	1	1	0
Japan	2	0	2
Nigeria	1	1	0
Norway	1	1	0
Philippines	2	0	2
Pakistan	1	1	0
Poland	3	2	1
Romania	1	0	1
Serbia	1	0	1
Russian Federation	4	0	4
Ukraine	1	1	0
United States of America	36	17	19
Uruguay	1	1	0
Vietnam	1	1	0
South Africa	5	3	2

34 countries

➤ Life on Mars

Have you ever wondered what it would be like to live on Mars? Your home would be a protected habitat without windows, you wouldn't be able to venture outside without a special spacesuit, and you would be totally separated from Mother Earth—if you needed a spare part or additional provisions, you'd have to wait at least a couple of years!

Enabling the ability to imagine life on Mars, so as to properly prepare for it, is the most important project the Mars One organization is currently engaged in.

Mars One is in the process of creating several Earth-based simulation outposts for training, technology tryouts, and evaluation, which will serve as the epicenters for the mission, both on Earth and on Mars. On Earth they will provide evaluation and training; on Mars, the final model will provide a safe home for the crew.

Mars One outpost

The general architectural layout and the interior design are some of the most important aspects of the outpost. These features will not only provide general safety, but also create a comfortable and enjoyable living environment.

The outpost design includes six assembled lander modules and two additional inflatables comprising living quarters, private areas, food production, life support systems, surface access, recreational areas, mission operations, life science, and much more. The inflatables will contain the main area of the settlement, providing approximately 200 square meters (about 2,152 square feet) for daily living and food production. Two centered lander modules will, when on Mars, provide access technology to the Martian surface, and the four remaining lander modules will mainly contain subsystems supporting the entire outpost.

A great deal of flexibility will be integrated into the interior design, making changes possible for special occasions or new situations in life. The outpost is also designed to expand as more astronauts arrive, creating more living space and ever-changing environments for the permanent settlement.

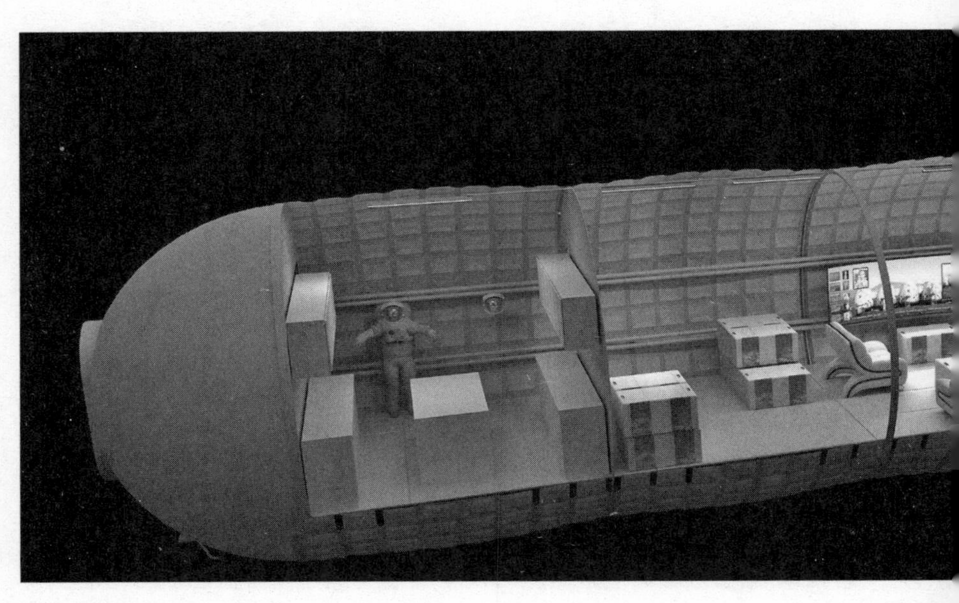

Mars One living unit

Mars One is working toward a production method that allows for easy container shipping, easy assembling techniques, and potential modular parts replacements. All of the modules will be rigid—even the "inflatable" volumes—and the interior elements will comply with the limits of allowed mass and volumes that can be transported and later be "unfolded" inside the outpost module shell.

Our astronauts will be settling on Mars indefinitely. It's not feasible to send water, oxygen, and food to them from Earth: They will have to produce these themselves. On Mars, water can be extracted from the soil, and the rover, which will be launched in 2022, will select the location for the settlement primarily based on the water content in the soil. In contrast to the scientific rovers dispatched to Mars to date, the Mars One rover's tasks will be focused more on utility—the deployment and maintenance of the human settlement on Mars. Water extraction will be performed by the outpost's life support units; the next utility rover, which will be launched in 2024, will deposit soil into a water extractor within the life support units, where the soil will be heated until the water within it evaporates. The evaporated water will be condensed and

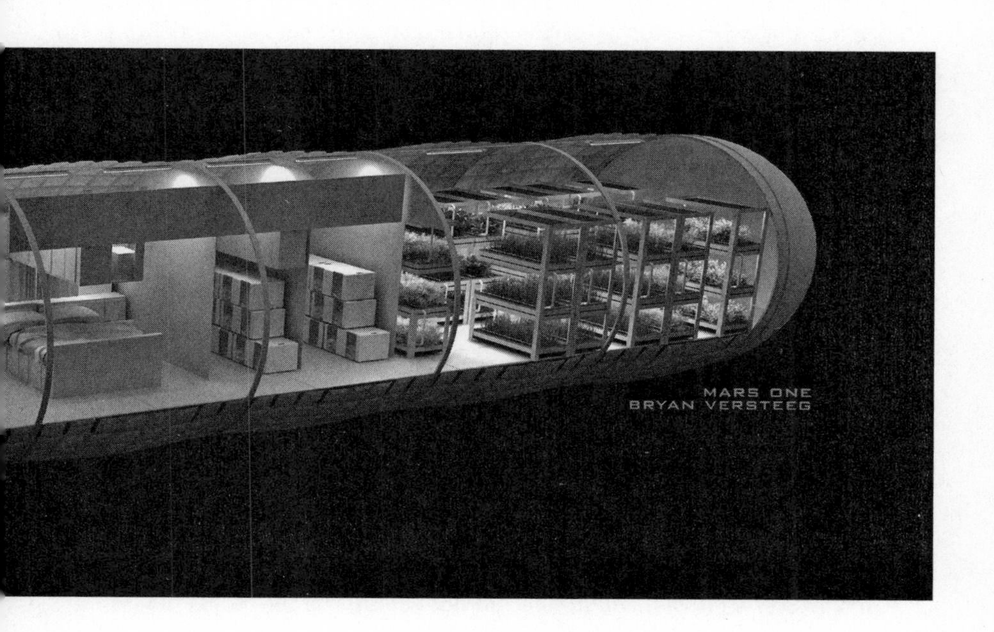

stored, the dry soil expelled, and the process continually repeated for daily use as well as to maintain reserves.

About 1,500 liters (about 396 gallons) of total reserve water will be stored in each life support unit; these reserves will be consumed primarily during periods of low power availability, for example at night, when energy must come from batteries rather than the outpost's solar panels, and in dust storms, during which noncritical systems will be shut down or restricted to save energy. (Because electrical power is so critical for the survival of the crew, extra safety margins will be built into the power usage profiles for dust storms.) The astronauts will have enough water stored for fifteen days of normal water usage (fifty liters, or thirty-two gallons, per day per person), and for one hundred fifty days if usage is limited. The water will also be recycled, which takes much less energy than extracting it from the Martian soil. Only water that cannot be recycled will be replaced by water extracted from the soil.

Oxygen can be produced by splitting water into its constituent parts, hydrogen and oxygen. The oxygen will be used to provide a breathable atmosphere in the living units, with enough oxygen for sixty days stored in reserve for times when there is less power available (again, at night and during dust storms). The second major component of the living units' atmosphere, nitrogen, will be extracted directly from the Martian atmosphere by the life support unit.

When the astronauts land on Mars, there will be storable food from Earth waiting for them to use, thanks to the 2024 unmanned supply mission. The storable food from Earth will serve only as emergency rations; astronauts will try to produce and eat as much fresh food on Mars as possible. It is likely that algae and insects will also be part of the Mars settlers' diet.

Food production will occur indoors under artificial lighting. In total, there will be approximately 80 square meters (about 861 square feet) available for plant growth in the original habitat, and the first crew will also be able to use the habitat planned for the second crew to grow food until the second crew arrives, as the hardware for the second crew's habitat will land only a few weeks after the first crew does. A thick layer of Martian soil on top of the inflatable habitat will protect the plants (and the astronauts) from radiation. Carbon dioxide for the plants is available from the Mars atmosphere, and water will be available through recycling and the soil on Mars. Nutrients for the plants could come from recycling human waste or be imported from Earth. Any plant production surplus will be stored as emergency rations, and nonedible parts of the plants will be recycled (e.g., to feed worms or into raw material for a 3-D printer). The aim is for the colony to be able to exist independent of any food it receives from Earth. But there will always be enough emergency rations in storage, locally produced or from Earth, for the crews in residence to survive on until the next supply mission arrives.

Crew members will have access to medical equipment on Mars and during the journey there. This will be needed to treat the most common injuries and illnesses, and all four astronauts in each crew will have received comprehensive medical training. However, the facilities on Mars will be more limited than those of a modern hospital. Due to weight limitations of the rockets and rocket launch capabilities, big, heavy equipment won't be present in the settlement for the first few years. Certain conditions (such as those requiring organ transplants, open-heart surgery, dialysis, and teeth implants) will be more difficult

Mars One living unit greenhouse

or even impossible to treat on Mars. Subsequent years will see more advanced medical technology make its way to Mars to allow for more complex care.

During their working hours, our astronauts will be busy performing three main tasks: settlement construction, maintenance, and research. The members of the first crew in particular will need to devote a lot of time toward the construction of the settlement, to make their new home into a comfortable place to live. They will install the corridors between the landers, deploy extra solar panels, and install equipment, such as greenhouses, inside the habitat. They will spend time on the crops and food preparation. They will also prepare the hardware for the second crew; the second crew's hardware will be delivered with the first crew of astronauts. Also, as early as possible, Mars One will try to supply the settlement with methodologies to produce additional living space from mostly Martian materials so as to significantly expand the settlement. Our goal is to enable settlers to construct a dome 10 meters high and 50 meters in diameter (about 33 feet high and 164 feet in diameter), which will provide a fairly spacious environment by Martian standards in which to relax, have a pond, and potentially grow small trees. This area will make Mars a much nicer place to spend time.

Maintenance will be crucial to ensure long-term functionality of all systems. The astronauts' lives depend on the technology present in the settlement, so all these systems need to be checked and maintained regularly.

Finally, research is also an important tool on Mars, and will be especially so once the settlement is fully operational. What is the history of the planet Mars? Did Mars have a long wet period, or just a few wet years every now and then? When did the dramatic climate change take place? Is there life on Mars now? The astronauts will do their own research and also collect data for other researchers and transmit it to Earth.

Our astronauts will also find time to relax. They will be able to do most of the indoor activities that people can do on Earth: read, play games, write, paint, work out in the gym, watch TV, use the internet, contact friends at home, and so on. There will be some communication and media limitations because of time delays resulting from the distance between Earth and Mars: They will have to request the movies or news broadcasts they want to see in advance. If an astronaut would like to watch the Super Bowl, he or she can request it, and it would be uploaded to the server on Mars. There will always be a time delay of at least three minutes, so the people on Mars will know who won the game a few minutes after the people on Earth. However, because any other communications from Earth would also run on the same delay, chances are this won't spoil the Martians' enjoyment of our "Earth sports." Easy internet access will be limited to their preferred sites, which will be constantly updated on the local Mars web server. Other websites will take between 6 and 45 minutes to appear on their screen—first 3–22 minutes for their click to reach Earth and then another 3–22 minutes for the website data to reach Mars. Contacting friends at home will be possible by video, voice, or email, but real-time dialogue is not possible, because of the time delay.

Living on Mars cannot be considered entirely risk-free, in particular during the first few years. There are a number of elements that could pose problems: An essential component of the settlement could

Mars One living unit relaxation area

be negatively affected by unforeseen events, and if an astronaut's Mars Suit were to become seriously damaged during an extra-habitat mission, there is a chance he or she might not survive. Also, as mentioned, certain medical conditions are not treatable on Mars. Obviously, Mars One will extensively examine and trial-run all elements of the mission beforehand to preempt any mishaps—especially the settlement's critical elements.

Overall, living on Mars is comparable to getting by somewhere like Antarctica and the International Space Station, and provides similar challenges—except that the South Pole has a number of very advanced large research stations boasting a great number of modern facilities that provide a good quality of life, and the International Space Station provides the opportunity to get back to Earth and the opportunity for constant supply.

These details describe the basic conditions of life on Mars, according to Mars One's plans; the "out of this world" essays that follow take this vision further, engulfing you in possible scenarios of daily Martian life.

First, however, to provide a complete picture of life on Mars, we must consider the sometimes very frustrating legal aspects of colonizing a planet. Narayan Prasad is an Erasmus Mundus SpaceMaster graduate and an EGIDE scholar (French Embassy Scholarship) with a master's in space technology (Sweden) and a master's in space techniques and instrumentation (France), with previous experience with the Indian Institute of Astrophysics and the German Aerospace Center (DLR). He is currently curator of the online publication NewSpace India.

His essay "The Politics and Law of Settling Mars: The Need for Change" brings us, however temporarily, back down to Earth with an uncomfortable thud, as he forces us to face the challenging political and legal consequences associated with creating the first interplanetary settlement outside Mother Earth. Prasad discusses international space law, with themes such as legal agreements needed by the international community on governance of interplanetary exploration and the need for such agreements on the subject of creating a settlement on Mars. And lastly we are confronted with the emergence of NewSpace activities and the ethical issues and potential legal barriers that could be raised with *regard to a private initiative launching a spaceship to Mars; but these issues will surely be dealt with when the time draws near!*

Next, Vincent Hyman, *who has written a number of articles for the Mars Exchange blog based on interviews with advisors to the mission, takes us through an imaginary day on Mars in "To Build a Bicycle: A Sol in a Settler's Life." Well—not quite fantasy—because the scenarios he depicts and the story he narrates could well be likely. As you'll see, as you enter the lives of his characters, who could well be the first settlers landed on a Mars One spaceship, carrying out a daily work routine on Mars is exciting and challenging, every escapade outside the habitat reminiscent of Captain Nemo's crew leaving the* Nautilus *sub deep underwater, protected by a cumbersome diving suit and with a limited supply of oxygen.*

Living on Mars is not all work, however—there will be time for leisure and for the settlers to do the things they always wanted to do, or perhaps never could have dreamed of doing. In "A Canvas as Big as a World," science fiction and fantasy author Robert Jeschonek *draws on his long history of fascination with travel to Mars to imagine how the first settlers on Martian soil will spend their leisure time, once the early busy work of setting up the habitat is completed—activities ranging from sports to experimental science to a general unleashing of "interplanetary creativity" remarkable enough to inspire two worlds.*

—The Editors

What are the political and legal complexities of colonizing Mars?

THE POLITICS AND LAW OF SETTLING MARS

The Need for Change

NARAYAN PRASAD

The launch of Sputnik 1 in 1957, the first man-made object to go into orbit, not only broke a technological barrier in humankind's ability to go beyond the boundaries of Earth's atmosphere, but also had significant legal and political implications. In particular, it established an international principle of "open skies," meaning that no nation can claim the space above the atmosphere as part of its air space, and so forbid satellites from crossing overhead without its permission.

Apart from its importance to space exploration, the de facto open skies policy had significant strategic and intelligence implications for a world locked into a Cold War, with spy satellites becoming a driving priority for both the Americans and Soviets. Consequently, in 1959, there were important political reasons for the General Assembly of the

United Nations to set up the Committee on the Peaceful Uses of Outer Space (COPUOS) to govern the exploration, development, security, and use of space for the benefit of all humanity and create a forum for reviewing legal problems arising from outer space exploration. The COPUOS acted as a foundation for the international community (IC) to debate and discuss the legal complexities involved, eventually leading to the creation of five major UN treaties (along with a number of lesser agreements relating to matters involving things like communication and meteorological satellites). These treaties provide a framework for international cooperation and coordination in the exploration of outer space, the most important of which is the Outer Space Treaty, which came into force as binding international law in 1967. The Outer Space Treaty forbids any nation on Earth from claiming sovereignty over any celestial object.

"Mars is like a time capsule, a billion-year-old record of ages long since past, bearing secrets for those brave enough to seek them out. We may not know the past of Mars, but we do know how the next chapter begins, because the story of Mars is far from over. After a billion years of isolation, we now wield the pen of history. This is where everything changes. Because, you see, going to Mars is so much more than just a mission of discovery; it's about how we cope in a crisis, it's about how far a dream can carry those who believe, it's about what it means to be human. I have no doubt that what we discover on Mars will send ripples of progress back to Earth: We will see the first truly novel twenty-first-century society born; we will see a scientific revolution in our knowledge of the universe."

—Mars One applicant

The last of the major treaties proposed to govern the activities of the ICs on the moon and other celestial bodies is the Moon Agreement, adopted by the UN General Assembly more than three decades ago, in 1979. Although there have been no concrete international political and legal discussions regarding settlement on Mars, the Moon Agreement offers an extension to Mars. It provides elaboration on and reaffirmation of many of the provisions of the Outer Space Treaty. The Agreement primarily provides that the moon and other celestial bodies be used exclusively for peaceful purposes without disrupting their environments, and that the United Nations should be informed of the location and purpose of any station established on those bodies.

Most spacefaring nations, including the United States, the Russian Federation (former Soviet Union), and the People's Republic of China, have neither signed, acceded to, nor ratified the Moon Agreement. (One of the primary reasons may well be one of the most controversial clauses of the Agreement, which provides that the moon and its natural resources are the common heritage of humankind and that an international regime should be established to govern the exploitation of such resources when such exploitation is about to become feasible.) Therefore, the general opinion is that the treaty is a failure because it has not been ratified by any IC that engages in self-launched, manned space exploration.

That no serious attempts have been made to either rescue the Moon Agreement from the legal doldrums or propose a replacement is likely due to the cooling of the space race following the landing of Apollo 11, and the novelty of various firsts in outer space wearing away. Consequently, the extent of political interest in space has been subdued, even among the leaders of space exploration. In addition, the focus of space activities has shifted away from purely national government programs, at least in the developed world. (The last decade has also witnessed the developing nations making substantial investments to match the historical achievements made by the leaders of the space race, although their progress is still somewhat limited.)

The last three decades have seen the emergence of global services based on satellites and more private industries delivering them internationally. Established spacefaring nations have been actively supporting the commercialization of technologies for gaining a larger international market share in global services. Much of this private activity has traditionally been either in the service of government programs—such as for launching spy, science, or weather satellites—or to support well-established industries on Earth, such as telecommunications satellites. However, the emergence of the "NewSpace" movement during the last decade is moving private enterprise toward engaging in areas very different than it has previously, including space tourism, development of swarms of satellites, asteroid mining, and so forth.

Traditionally, many of the political and legal issues surrounding the activities in outer space have been discussed in the context of being led by international communities, with virtually all aspects of governance and codes of conduct also held within their purview. With the emergence of NewSpace, it is important to acknowledge the change of space activities from an IC-driven model to one that is driven by a private person or enterprise, or other nongovernmental actor. This leaves it somewhat outside the scope of the UN treaties and agreements, which are built on the paradigm of regulating the activities of national governments, or in limiting the impacts of space snafus to those living on Earth (such as the rules that apply when a spacecraft or satellite crashes somewhere on the surface).

When it comes to the involvement of private and nongovernmental entities, the current legal vacuum will have a significant influence on the political and legal complexities involved in working toward creating a self-sustaining settlement on Mars. For example, should the UN lead the discussions? A single IC? What about a completely nongovernmental, private approach? There could also be a multi-IC/multi-private enterprise or an IC-backed private-enterprise effort.

Although some of these organizational approaches are less likely to be realized than others, each possibility considered would have profound effects on the discussions that the political and legal apparatus of each IC would bring to any decision-making table. The organizational approach, therefore, has a fundamental impact on the issues of governance and overseeing of activities for creating a settlement on Mars. Furthermore, although the current construct of international space law necessitates that international communities oversee and authorize the activities of all actors under their jurisdictions, several issues of interest to private enterprise and nongovernmental actors remain to be clarified under the ambit of international space law. For example, the creation of a Martian settlement includes exploitation of resources, such as valuable new minerals, for their eventual commercialization and possible transport back to Earth, or transforming the very environment of Mars for the creation of a self-sustainable human habitat or for tourism. The legalities of such actions remain hazy under the current international space law.

The prohibition of nations in establishing legal claims on nonterrestrial territory is a major clause of the Outer Space Treaty, which renders that "the outer space, including the Moon and other celestial bodies are not subject to national appropriation by claim of sovereignty, or by means of use or occupation, or by any other means." Similarly, the same treaty mandates that the conduct in exploration of outer space should avoid its harmful contamination. However, there is no clear definition of either outer space or a celestial body given in the treaties and therefore no consensus on the definition among the international communities pursuing space activities. Given the current stance of international space law, any activities pursued toward the creation of a Martian settlement may be perceived as de facto appropriations or not, based on the political alignments of ICs against such actors. There is a need for ICs to reengage in serious multilateral discussions to resolve such issues.

Such multilateral discussions are born when one IC initiates the process of working toward a resolution and building a critical mass of ICs who concur in their agreement, based on their common faith in recognizing the need for such a resolution. However, this process in the realm of international law is long and tedious. In addition, treaties and agreements that result from such a consensus-driven process can also suffer from the lack of an international enforcement body or mechanism against violations of international space law.

These problems, together with the fact that activities beyond Earth's orbit are limited to relatively few actors, point to a situation where ICs may want to address the legal issues involved via the vehicle of national legislation.

Such a route is already gaining some traction within the United States, where the legislators passed the SPACE Act of 2015 (H.R. 2262), in May 2015, to facilitate the commercial exploration and utilization of asteroids in accordance with existing international obligations. The bill addresses some of the issues of appropriations, licensing outer space mining, jurisdiction over outer space resources, and commercial property rights for companies that wish to mine asteroids for resources.

The increase in the number of parties attempting such endeavors will result in the emergence of a competitive market, with the actors channeling major resources into resolving technological challenges. Eventually, those ICs that have jurisdiction over such parties may evolve internal regulatory bodies with specified roles covering authorization, insurance, liabilities, export control, tax rationalization, safety standards, and so forth.

Nevertheless, there is a need for political engagement of ICs at a multilateral level to create a consensus in extending the current international legal framework into one that shall allow for reaping commercial benefits in return for the investments made by private individuals and organizations.

Ideally, we would establish a modus operandi of governance that would allow the exercise of economic, political, and administrative authority in order to manage affairs at all levels in the exploration of Mars, the creation of a self-sufficient base there, and its eventual settlement. The aspects of governance need to comprise the mechanisms, processes, and institutions through which the ICs, their citizens, and groups articulate their interests, exercise their legal rights, meet their obligations, and mediate their differences.

A long-term vision of a fully functional Martian settlement would encompass governance of every institution and organization in the society, right from the individual to the family to the entire community. It should embrace all methods that such a settlement should adopt to distribute power and manage resources and problems within its local environment.

In the light of such a long-term vision, existing terrestrial ICs need to work toward establishing a strong legal foundation for such governance. This will require them to come to a consensus on the form and nature of governance, including how power will be separated among different institutions. The ICs then must agree on the rights and duties of those pursuing activities on Mars and authorizing private enterprises or nongovernmental actors to also do so. This top-down approach in setting up a legal foundation will then need to reach the point of detailing aspects of good governance of Martian settlements.

A well-established framework that could act as a template for dealing with the plausible issues of a Martian settlement is the work of the United Nations Development Programs (UNDP), which identify a number of core characteristics as a measure of good governance, including the participation of the masses in decision-making; a fair and impartially enforced rule of law; transparency built on the free flow of information; a responsiveness to serve all stakeholders; consensual mediation of differing interests; equity to improve or maintain the well-being of stakeholders; effectiveness and efficiency of processes and institutions; accountability of all concerned stakeholders; and strategic broad and long-term visions for governance.

> "I admire the idea of creating a brand-new human settlement from the start with new rules and taboos. There are so many things you can do. No limits for our inspiration!"
>
> —Robert P. Schröder, Mars One applicant

Such a framework for governance should evolve based on the unique social, economic, and environmental aspects of the settlement. For example, some of the core issues in creating a self-sustaining ecosystem on Mars may well point to the idea of terraforming the planet. In such a scenario, there is need to establish a legal authorization and political decision-making chain that would allow such actions to be pursued. Similarly, the social aspects may involve providing the right of an individual to return to Earth.

Within the very nature of settlement—as opposed to simply scientific exploration, as occurs in Antarctica—lies an important ethical angle that is interweaved with politics and legalities. Modifying the environment on the planet in ways both large and small to suit sustainable human habitation will impose different moral weights on political decision-makers and in the construction of legal framework enabling such conduct than would occur with purely scientific exploration.

Unfortunately, it has been difficult to get terrestrial ICs to seriously consider the moral dimensions of activities in space—for example, the increasing space debris orbiting Earth. Although there have been some cases when acts of reckless creation of space debris by ICs have received widespread criticism, such as military-technology demonstrations of antisatellite weapon tests, the resolution of matters in cleaning up of the space debris has found no substantial traction in the political realm of the spacefaring nations.

This stalemate suggests that gaining a critical political mass toward a binding international resolution may well not occur until a significant escalation in the severity of the problem occurs (in this case, probably an event pushing us close to the Kessler syndrome, which would leave Earth's orbit a no-go area as cascading showers of debris made space travel too dangerous). There is need to preempt such a stalemate while establishing a settlement on Mars. The muscle of polity and jurisprudence must be built over a strong skeleton of morality when applied to the case of building a Martian settlement.

Preempting this stalemate, and resurrecting significant international political and legal engagement by international communities to resolve the legalities concerning activities such as asteroid mining or Martian colonization, is only likely to happen following concrete demonstrations of the technological and commercial viability of such activities.

Other likely scenarios for expedited international engagements could arise out of existential threats to human life on Earth due to climate change, a dangerous approach of an asteroid or comet, a global pandemic, and so forth. However, the birthing of an international political and legal engagement is more likely to arise out of the economic interests of ICs considering future returns than out of these more dramatic scenarios.

Traditionally, law has often lagged behind scientific and technological advances. With the fast-changing dynamics in the technological and economic landscape of nongovernmental institutions leading outer space activities, there is an emerging trend of such actors voluntarily declaring a code of conduct based on their social responsibility. For instance, several NewSpace satellites companies, such as Planet Labs and Spire, have announced their plans to ensure controlled reentry of their space assets despite no hard law binding them to do so.

Such measures not only provide scope for aligning their space activities to ensure long-term safety and sustainability, but also provide an opportunity to boost confidence among investors and other concerned observers of such missions.

Given the lack of political urgency among ICs to work toward an international consensus on the impending legal issues around creating a Martian settlement, there may well be an upward trend in the independent formulation of national space laws, and also more non-IC actors declaring their code of conduct for their activities in assuring space safety and sustainability, thereby creating a new de facto international political framework.

Narayan Prasad *is the cofounder of Dhruva Space, a NewSpace company based out of Bengaluru, India, building small satellites. He is an Erasmus Mundus SpaceMaster graduate with previous experience with the Indian Institute of Astrophysics and the DLR Institute of Space Systems, Germany. His interests cut across technology, law, and business aspects of space exploration and utilization.*

What will living conditions and quality of life be like on Mars?

TO BUILD A BICYCLE

A Sol in a Settler's Life

VINCENT HYMAN

"Di, there's a flare coming in. Not too big, but we need to get you back as soon as we can. You're going to have to abandon work on the drill today."

Di swiveled to gaze at the distant habitat behind her. A trail of boot prints and rover tread marks led to dusty mounds of red regolith that sheltered its domes. In places, the wind had already softened the crisp imprints of her lugged soles.

Ahead, the trail forked in three directions. Through butterscotch haze, she saw the drill site straight in front of her. Two closer, equidistant radiation shelters marked the ends of the other trails.

"Mei," she radioed, "I'm too far to turn home. I'll head to Shelter One."

Frustration, again. More than seventy days on Mars and finally they'd had time to set up the drill site to search for signs of life, past or present. After studying satellite photos, Mars One's science consultants had chosen the closest likely spot—a small basin with ancient signs of water erosion three klicks from the habitat.

Looking for life. If this had been *her* mission and hers alone, the search for Martian life would have been job one—right after pinching herself to be sure she'd arrived in one piece. The whole world was watching and waiting. Do we have neighbors? The tiniest life-form would change everything—more than Copernicus did by proving that Earth revolved around the sun, more than the first photo of Earth shot from space, more than the brash idea of settling Mars. Despite countless books and films featuring alien invasion, the truth was that humans were hungry to learn they weren't alone.

Last month, Eedris had led the weekly broadcast classroom chat and devoted it to the topic of prepping the drill site and searching for life. He'd opened with one of his playful poems, still stuck in her head:

A tiny germ
Might not be
The Martian friend you'd look to find
But as a sign
We're not alone
I'd welcome the microbial kind.

Eedris was a ham and a charmer. Always ready to spend time on things that weren't critical to survival yet critical to enjoying life. Even on Mars—*especially* on Mars—what was the point of surviving if you couldn't find joy? His humor made him a good med tech and a great teammate.

Of course, survival came first. As the team's senior engineer and glorified mechanic, Di had led them in establishing their habitat. Along with Saeed, Mei Vang, and Eedris, she had spent most of the past two months just getting the basics in place. More to do—so much more—but the folks back Home, especially their sponsors, needed a bit of magic now and then. Every member of the team was eager to deploy the drill and search for evidence of life a few meters down.

She stopped a moment to check in. "Mei, Shelter One is just minutes away now. I'll call in when I get there."

Di and Saeed had crafted two solar storm shelters so far, the beginnings of a planned network of radiation-safe havens that would enable longer excursions from base. Scraped up of loose stone and covered in regolith, the shelters were barely fit for cavemen but good enough to weather a radiation storm. "A romantic little culvert for two," Saeed had christened the first one. "We should put candles in there and a bottle of Georgian wine. Except the candles wouldn't burn and the bottle would freeze. Otherwise, perfection."

So . . . a few more minutes across the rust-red rock and dust to the shelter, then wait out the storm, and then I'll have to head back, she thought. *Not enough air left to work on the drill site today.*

Di bounded toward her stone culvert. She loved running on Mars, and she welcomed the time away from the noisy habitat and the prying gaze of the video cameras there. The solar storm would provide precious moments alone.

"For the longest time, all I ever wanted to be was an astronaut. I wanted to sail through the inky black unknown and land, explore, and survive on a place like Mars."

—Mars One applicant

"Di," Mei Vang's voice crackled. "I can hear you breathing hard. That means everyone can hear you. Or *will* hear you when the transmission reaches Home. Your respiration's up. Mission rules: No running. No trotting. No jumping. At least until we're better settled in and get the go-ahead from Control. Walk to the shelter and wait. No hurry, and no need to scare anyone. No one wants to see you take a tumble. And *please* turn your video feed back on."

Of the Four, only Mei naturally balanced her sense of duty to Home, her life on camera, and her need for personal space. Was it cultural? Big extended family? Eedris loved to make faces for the cameras. But this life-in-a-fishbowl adjustment had been hardest for Di and Saeed. "Like living in a zoo," Saeed had said.

Di could understand Home's fascination with Mars, the mission, the tech, the settlement, and the science. What she couldn't fathom was the public's enthrallment with the Four as individuals. It had stunned her to learn that millions were tuning into MarsTube to "dine with the Martians." What was the fun in watching a delayed video feed of four people drinking weak tea and munching processed egg white, fiber, and protein bars?

Oddly, according to the news reports and commentary, the whole ritual was beginning to have a unifying effect back Home. When people in Antakya, Honolulu, New York, Reykjavik, Addis Ababa, Mexico City, Yap, Beijing, and Mars sipped tea together, the world felt smaller. Correction, the *solar system* felt smaller. She'd had no idea when she volunteered thirteen years earlier that her job duties would include drinking tea with billions, or that people would be commenting on her hair and speculating about her love life.

At breakfast, Mei always led a toast to Home, welcomed viewers, wished them well. *What an amazing young woman*, Di thought. A descendant of the Hmong mountain people of Laos, born in rural Wisconsin, Mei was a half-meter shorter than Di and Saeed, but she was arguably the toughest specimen on the team. Di, the oldest of the crew, felt like a wind-tossed poplar next to Mei, despite her youth and short stature.

Mei had become the public's sweetheart, and that was just

"I want to show humanity what we can do when we knock off our petty bickering and see one another as allies instead of strangers. One of the most unifying events in our species' history was watching Neil Armstrong walk on the moon. For those few moments, we weren't a planet of white people or black people or Jews or Muslims or Manchester United or Liverpool FC. We looked at the moon and said to ourselves, 'Yes, we did that.' And we weren't America or NASA or Neil, Buzz, and Mike. We were the whole world. We have forgotten what that feels like. I want to remind people of what we've done, of what we can still do, and that the sky is never, ever the limit."

—MARS ONE APPLICANT

---

fine with Di. They also loved Eedris—his hamming for the camera, his funny poems, his broad smile, and the way his great dark hands wagged about as he spoke. Saeed's nightly Martian weather reports were a huge hit. They'd been picked up by Al Jazeera, CNN, BBC, and hundreds of local networks—earning needed extra cash to help resupply the Four and build support for the next teams. "We're expecting a balmy twenty below Celsius today," he'd intone, waving his arm across a hand-scribbled map of Mars. "Winds up to two hundred kilometers an hour, but with the low pressure, that's like a mild breeze on Earth. A perfect day for some sailing, if we had an ocean, some oxygen, maybe a boat . . ."

Di was oldest of the Four. Calm, quiet, always the engineer; the public had dubbed her "Oma Di" a few years back. It was a salute to her Dutch heritage and her age. She *felt* neither old nor grandmotherly, but the name stuck, and eventually her teammates had begun to tease her with "Oma Di." *Oh, well.* That was her public face, and what she thought about it mattered little. This was, after all, not just her venture. It was humanity's venture. Everything the first Martians did mattered.

Ignoring Mei, she leapt over a few small boulders and headed toward Shelter One. Two months in, and her newfound "superpowers" still made her giggle. All those times she'd flopped at the hurdles in gymnasium—athletic feats were never her forte—and now here she was at age fifty-nine, leaping like Wonder Woman. On Mars, at least, she was better than the best of Earth's Olympians. If only the spacesuit were more flexible. Maybe some sort of heated, flex material. A compression system of sorts? Her engineer's brain charged ahead of her, and she shut it down to refocus on reaching the shelter.

There it was at last. A clumsy bend at the knee got her ready to enter. "Mei, I'm going in, and I'll want my silence. Check in in ten minutes, okay?"

Time, thought Di. Time alone, to think.

What a day! She'd skipped Breakfast-with-Earth to allow time for her extra-habitat excursion prep and med check. She used the morning to review her last vid call—her mother was not doing well and spoke little. Meanwhile, her aunts wasted valuable airtime, talking about everything from the latest royal birth to her kid cousin's latest holographic tattoos. Their chatter spoke volumes by what it left out: Her mother, their sister, was fading. Di knew this was coming before she left Home. They'd made their peace and said their physical good-byes, and she was thrilled that her mother had held on long enough to see her youngest daughter surviving on Mars. It felt awful, though, to be so far away and know that her mother would soon be gone. Di could be with her on video feed for the last moments—sort of. With the time delay, her mother would be gone when Di found out. Or would she know the moment it happened? The engineer in Di said no, but the mystic in her craved yes. Small comfort that she had it better than settlers of centuries past, who often never heard from family once they set sail across the oceans.

She'd wasted a few minutes skimming some favorite web pages, downloaded the previous night. New photos from Facebook. Her latest spacesuit shot, the one with the habitat reflected in her gold visor, had gone viral. A few dozen clever folks had already photoshopped images from the Taj Mahal to their own homes to Mickey Mouse's face onto the reflective visor. It was fun, but there was too much of this stuff to wade through, no matter how clever. She'd have to get Mars One to adjust her filters, or get some kind of volunteer help sorting this stuff. Like her time, the data bandwidth was too precious.

Earth news was the usual mix. "Storms of the century" on the rise, New Orleans doused again, her beloved Netherlands investing trillions in seawalls and pumps and paying it all off by exporting the technology. More advances in organ printing, and there was a private email from their medical sponsor about shipping an organic printer along with the third round of settlers, due in six years. *Nice idea*, she thought, and one Eedris would welcome. But she'd rather get the necessaries for a closed-arc furnace so they could take advantage of all that iron out there. They had a lot of building to do. Of course, Mei wanted more Kevlar domes, and favored importing genetically modified bamboo sprouts to *grow* structural supports. No reason they couldn't run the numbers and try both options, with the right sponsors and money. Just keep those ratings up and the money will follow!

A team meeting had been next on the day's agenda. Eedris presented an update on the progress of the hydroponics system. They were a full two weeks behind: Aside from some sprouts, they had harvested no fresh greens yet. She missed those! Mei was still struggling with the hydroponics' water cycler. Saeed and Di had suggested several workarounds for a broken pump until they could get the 3-D printer back online and generate the needed part. For now, the best bet seemed to be manually pumping water to a gravity feed.

> "The question asked most by mankind is, *What am I doing here?* By nature, we long for meaning and purpose; our greatest fear is being forgotten. We want to leave a legacy. I can think of no greater story for my family to tell their children and grandchildren than the story that I left this planet to begin something great. Literally, to begin New Life."
>
> —Kirby Baumann,
> Mars One applicant

The weekly "live" chat with schools across Earth had come next. It was always the best time of the week. Nothing was really live, but the classrooms taped their questions live and the team's responses were, too—then it was all pieced together by Mars One and rolled out on MarsTube. Everyone loved these sessions. There were already close to 750 million followers. The kids were so energetic and inquisitive and surprising, and they often brought up questions adults were too embarrassed to ask.

The week's topic had been daily life on Mars, and the first question had come in from a class at a boys' school in Islamabad. "How do you go to the bathroom on Mars?" asked a wispy boy. Eedris mugged for the camera, Mei sighed and rolled her eyes, and Di and Saeed chuckled. Ten-year-old boys were the same 'round the world. Saeed explained the process, including the valuable water and waste reclamation systems.

Then a middle school class in Antwerp came online. "What does the habitat smell like?" asked a girl. "Does it smell like a locker room?" chimed in a slight, dark-eyed boy, and the class cheered.

"Truth be told," said Mei, "it probably smells pretty bad, but we can't smell much. It's . . . *our* smell, you know, so we don't notice it. Do you know what your own room smells like? It's the same here. You get used to the smell and then you don't smell it. Except in the hydroponics gallery! There, we can smell the fertilizer and the precious water and the new plants that are starting to come up. And those smells are . . ."

"They are *delicious*," broke in Eedris. "Like rain at the end of a hot summer day. Like the scent of your mother when she hugs you after a bad day at school. Like heaven. To breathe in that gallery is to inhale our history on Earth and exhale our future on Mars. That delicious scent is the closest thing we have to the smell of Home, and we all love to spend time there."

Now they heard from a class in Quebec. "How do you wake up in the morning?"

"Waking up is the best," said Saeed. "I change my wake-up sounds and projection-wall scenes every day. Sometimes I set the channel for the sound of rain. Sometimes it's the call of rock doves in Tehran. Sometimes it's old recordings from the Aref Ensemble or the Minnesota Orchestra. Sometimes it's a view of the Martian landscape and the sound of wind. Sometimes it's an owl hooting, or the wind in the pines, or waves breaking on the coast. I even have a recording of my mother telling me I'd better get a move on or I'll be late for school! I tried some Big Bang Metal once, but *that* didn't work out so well. I had the volume turned up too loud. My sleepy brain forgot I was on Mars and I jumped so hard I smacked my head on the lamp. Almost hit the ceiling! Mars One's psych staff was a little suspicious about my black eye. They asked us if we were fighting. From now on, I'm reserving the heavy metal for my fitness routines!"

Next had been a teen girl from Kyoto. "Don't you miss home? There are two men and two women. Don't you get . . . lonely?" Then she blushed. "Aren't you in love? Everyone all over gossips about you. How do you deal with . . . all of *that stuff*?"

Eedris stepped up. "Look," he said. "This is a normal question. Something the whole world is curious about. Our personal lives are still ours, and we will keep them that way. That is to say, we will keep private things private, just as you would. We know that there will always be gossip and questions. Questions of intimacy and emotions are things that we *do* talk about among the four of us in private, but have agreed as a team we will not share with you. And some of us have close attachments on Earth who we will love always, and we talk and email with them every day.

"People also are always curious about having babies on Mars. There will come a time when people have babies here. But that is not in our near future. For safety reasons, a pregnant woman cannot leave the habitat. So if one of us becomes pregnant, then we are reduced to three who can leave the habitat. And if one of those is ill, then we are two. On Earth, childbirth is fraught with health risks. Raising a child, especially during the times when the child needs constant attention and supervision, also changes the function of the team. So these things all play a role during this critical time. But," Eedris continued, "at some point, there will be more settlers here. If all goes as planned, within six years there will be twelve of us, and then perhaps more and more. When space and safety and systems are more reliable, children will be a part of our future on Mars."

"And really it is not good-bye, it just means that we won't have physical contact again. We can accept that. It's only about six light-minutes away after all."
—Mars One applicant

Mei added, "You also asked about whether we're lonely. It's not like we're alone. There are only four of us, true. And the thin, toxic atmosphere outside the habitat would destroy us in moments. The landscape is barren. The average temperatures are colder than my home in northern Wisconsin on the worst day in winter. And we can't just have a conversation with people back home the way you do; it takes time to send messages back and forth between our worlds. So that sounds lonely, right? But we have *so many* connections at Home—we have friends all over Earth, and we have breakfast with millions of people every day. We certainly are more in touch with family, friends, and the whole world than any settlers or explorers in the history of humanity. Probably we are connected to more people than *anyone* in history. We feel fortunate."

Di responded next. "We trained for this for ten years. We aren't a family. We aren't really like anything most people are familiar with. Well, maybe we're a little like a really good basketball team or jazz quartet, only more flexible. We cross-trained to do all kinds of things well. We had to prove we were the best team on Earth in order to make it here. After ten years of winning and losing together, we know one another very well. Of course we are close to one another. We each know when to support, when to lead, when to follow, when to listen, when to fight, when to shut up, when to stand in each other's way, and when to get out of someone's way."

"Oma Di has spoken," Saeed joked. "And I, for one, am going to stay out of her way." He folded his arms and bowed his head in mock deference.

Next a preschool class in Santiago. "Do you ride bicycles on Mars?" asked a ginger-haired boy. "I watched you in space with your pedal exerciser. Can you ride around out there on Mars, too?"

Everyone had burst out laughing. *Bicycling on Mars?* Di stepped up for that one. "We don't have a bicycle," she said. "The pedalciser was left on the space module and it's orbiting the sun forever. To get around, we walk, or we have Fido, our rover. I'm not sure if we *could* bicycle on Mars. You can't bend your knees too well in a spacesuit. Then I don't know if the wheels could grab. Then there's the problem of the fine dust clogging the gears. The greater risks of falling. And the air pressure in the tires . . ." Before she knew, it her mind strayed into an ever-expanding tree of linked problems and solutions. Eedris had touched her arm to bring her back.

"Bicycling on Mars sounds like a fun idea, but for now that's impossible," Eedris said, and his big grin, slight bow, and sweeping hands signaled a theatrical close to the session. "And that wraps it up for this week. Thank you, everyone, for your questions! Next week we will be taking questions about our garden. Mei and I will send a video tour tomorrow. After you see the tour, video your questions and send them to Mars One by Monday. And remember, we rely on our sponsors. So ask your parents to help out however they can."

Di's reverie was interrupted by the crackling of the communicator. "Di, looks like you're clear. Time to head back. We'll have to reschedule your work at the drill site with Mars One Control."

A twenty-five minute walk back. Time enough to think about . . . bicycles. She could not shake the image. Could they build one? With what? Why? But then, *Why not?*

It could be done, she saw. Wouldn't need to be a crank pedal. A treadle or elliptical system? Friction or direct drive. Sealed. Some spare rover parts. Three wheels better. The problems and solutions were buzzing fast.

"We as a species have barely begun to even understand ourselves, but we are trying hard to find out everything we can about us and the world we exist in. The chance to explore a completely alien world, to start a new society from scratch, to take on the everyday challenges this will bring, to find something new and rediscover the long-forgotten old is what drives me forward."

—Mars One applicant

Leave it to a five-year-old, she thought. Go for a bike ride on Mars, with everything there was to do? She remembered her first tentative time on her own bike, her older sister guiding her. The blurry surroundings. Falling, getting back on, falling again. Balancing. Getting it. Riding more, riding faster. And finally, the marvelous sense of freedom and independence. The capacity to venture far from home.

Riding a bicycle on Mars. Silly, to put the time and resources into such a scheme. But then, what did it say when a five-year-old was starting to see their lives on Mars as normal? What better reason than to show that they *were* normal?

"Mei," Di called in. "I'm almost back. It'll take a good forty till I'm checked and cleared. But let's get everyone together to talk while we wait. I can't get that little kid from Santiago out of my head."

"We're way ahead of you. Saeed's already sketched some plans," Mei said. "Eedris knew where you were going before *you* did. It's crazy, but no less so than coming here."

"It's not crazy," Di said. "It's audacious. It's what we humans do."

"You know that when we tell Home what we're up to, the chatter about wasting time will start," Mei cautioned.

Who says they need to know? Di thought, but said nothing.

It would be their surprise.

Vincent Hyman *is a freelance writer and book editor specializing in mental health, nonprofit management, science, and addictions. A Mars One applicant, he contributes to the mission by interviewing their expert advisors and writing articles for Mars One Exchange. He has directed publishing operations for nonprofit consultancies at Fieldstone Alliance and Amherst H. Wilder Foundation and worked in public relations and marketing communications. He lives in St. Paul, Minnesota. To learn more, see www.vincehyman.com.*

A CANVAS AS BIG AS A WORLD

ROBERT T. JESCHONEK

Imagine it's the late twenty-first century. Mars comes into view in the night sky, and telescopes on Earth focus on its dusty red surface. The reflected light of the sun reveals an image from millions of kilometers away, an amazing and historic vision witnessed by billions of breathless viewers.

A human face, framed by what looks like the headdress of an ancient Egyptian pharaoh, can be seen on the plain of Utopia Planitia, etched in the red dirt over hundreds of square meters. Human hands have carved a figure into the Martian landscape that, like the Nazca Lines of Peru, can best be seen from a great elevation.

Decades ago, humans thought they saw just such a face on Mars. It turned out to be a trick of the light . . . but now a great face exists on Mars after all. Human settlers have brought the legend to life.

For the first time, a human artwork can be seen from Earth on the surface of another world. Mars, with its vast reaches of empty land, has made this possible. Human inhabitants, the descendants of the original Mars One settlers, have made it a reality. Like so many other explorers before them, they have found inspiration in a new world of fresh possibilities.

They have put their leisure time to use channeling a force quite unlike any other yet encountered in space and time: human creativity.

It's fun to imagine the leaps of expression and achievement that might one day be possible for human settlers on Mars. When boundless creativity meets vast, open landscapes, the possibilities will be limitless.

But in the early days of Mars colonization, leisure-time activities will likely be more modest. Drawing Nazca-style figures on the sands of Mars just won't be an option.

In the first days of colonization, leisure time itself will be practically nonexistent. The Mars One settlers will simply be too busy with the work it will take to ensure day-to-day survival in a hostile environment. Unbreathable air, bitterly cold temperatures, abundant solar radiation, and a lack of liquid surface water will force the settlers to live mostly indoors, in small, climate-controlled habitats.

Preparing these living quarters for habitation will be at the top of the settlers' to-do list when they first land on Mars. Though inflatable components for the habitats will have been delivered to the surface of Mars before the settlers' arrival, they will still have to assemble the components into a functional settlement. It will take time to get everything up and running, from the power supply to the internal atmosphere, heating system, food production unit, and water recyclers. It won't be easy or quick, especially since the outdoor work will have to be done by personnel wearing Mars Suits—bulky environmental suits similar to those worn by astronauts on the moon or in space.

Even after that first flurry of activity, the settlers will be fully occupied with the maintenance of the settlement. They will dedicate time to the long-term, ongoing project of improving their living conditions, such as the way they extract water from the Martian soil; produce and prepare food; and recycle and reuse fluids, refuse, and other materials. They will strive to find and implement alternative energy sources and come up with innovations in everything from mechanical repair to medical treatment.

Exploration will also be a priority for the time left over when basic needs have been met. Eventually, settlers will range outward from their home base, exploring the Martian surface and conducting scientific tests en route. To expand their reach, they will need to set up safety stations where they can refuel or repair their rover vehicles, pick up supplies from a cache, take refuge in case of emergency, and signal for assistance as needed.

The Sadness of Mars

Those first humans on Mars will be millions of kilometers from home with little or no hope of return. They will be permanently separated from loved ones, without real-time audio or video contact. (Due to the distance from Earth, the best-case connection will have a delay of seven minutes each way.) And they will be confined, for the most part, to limited indoor living spaces. Settlers will only be able to go outside for a few hours at a time, wearing Mars Suits; otherwise, their daily lives will play out in an enclosed space of about two hundred square meters, with limited privacy or opportunities for a change of scenery.

"Of course, conditions will be very hard. But haven't we learned to survive in extreme circumstances since ancient times? Curiosity has always been the essence of human existence on Earth. Mankind has always tried to find out what's on the other side of the mountain. This journey to Mars will be the ultimate challenge and I want to take part in it."

—Mars One applicant

Homesickness, loneliness, and confinement, plus the pressures of survival on a world where even the smallest necessities must be eked out of a hardscrabble environment with little margin for error. Could there be a better breeding ground for depression?

Once it takes root, depression could be as deadly to the health of the community as the thin atmosphere, brutally cold temperatures, and solar radiation on the Martian surface. Between the daily work necessary for survival and long-term projects related to process improvement and exploration (not to mention sleep), much of the settlers' time will be spoken for. True leisure time will be in short supply but also one of the most precious commodities on Mars, vital to maintaining the mental health of the settlers. How they spend their leisure time, under what are sure to be very stressful conditions, will be just as important to the mission as how they grow their crops or maintain their inflatable habitats.

Binge, Baby, Binge

What will the Mars One settlers do to make the most of their free time and fend off depression? NASA's HI-SEAS project has provided a preview.

HI-SEAS—the Hawaii Space Exploration Analog and Simulation project—was developed to simulate the conditions of living in the Mars One settlement. Groups of volunteers have spent months living in a domed habitat on the slope of Mauna Loa on the Big Island of Hawaii. Like the Mars One colonists, they've mostly been confined to their indoor habitat, going outside only for limited periods, wearing bulky environmental suits.

The HI-SEAS teams have performed duties much like those that the Mars One crew will undertake, and their contact with "home" has been extremely limited. Delays have been imposed on communications and the internet, preventing real-time interaction beyond the world of the dome.

> "For me, Mars One represents the opportunity to explore the truly unknown, much like the explorers of old days; to find and learn new things, whatever they might be."
> —Mars One applicant

So how have the HI-SEAS folks put their free time to use? They've spent a lot of time exercising to maintain muscle tone and bone density, which is something the Mars One crew will have to do in the lower-gravity Martian environment (which is approximately 38 percent of Earth's normal gravity). But maybe we shouldn't count exercise as a leisure activity, as it won't really be optional if the settlers want to stay healthy.

So let's move on to more leisurely leisure activities. According to Zak Wilson, a participant in the 2014 to 2015 phase of the HI-SEAS project, the most popular leisure-time activities inside the solar-powered dome include watching prerecorded TV shows and movies, playing board games, and reading.

Makes perfect sense, doesn't it? What better time to catch up on your favorite shows and books than when you're cooped up in a habitat on Mars for the rest of your life? The modern pastimes of binge-viewing, binge-reading, and binge-gaming will be the ideal leisure-time activities for the Mars One settlers (in between working hard to survive, performing scientific experiments, exploring the surface, etc.).

But man and woman do not live by bingeing alone. Won't the settlers eventually get bored with videos, books, and games? What happens when these forms of entertainment are no longer enough?

They can always put their spare time and energy into entertaining the folks back home.

The Greatest Show on Mars

The Mars One astronauts will live out their lives farther away from Earth than any humans in history. They will be physically remote from the rest of humanity and unable to communicate in real time. Yet their lives are sure to be among the most closely watched of all time. In the early days, especially, their grand adventure is certain to fascinate the billions of viewers glued to the latest video and audio feeds from the Red Planet.

To help cover the costs of colonizing Mars, the Mars One organization plans to provide content for a reality-show broadcast to Earth audiences. The settlers will be the stars of this show, shooting video of their activities and interactions and transmitting it back to Earth.

Could there possibly be a more surefire smash hit? People around the world will be dying to witness every new development at the history-making settlement, from the greatest to the smallest. And providing access will make the settlers feel more connected to Earth, giving them something to do that is mission-oriented while also being fun.

Shooting segments for the show will be the perfect leisure-time activity; a great outlet for self-expression and performance that isn't a matter of life and death. The settlers will be able to act as serious or silly as they like . . . stage stunts and pranks for the camera . . . and bare their souls to the folks back home. It will certainly provide a vital distraction from the constraints and dangers of life on Mars and a meaningful way to frame the presentation of the experience for Earthbound humanity.

> "For my entire life I've looked at the sky in wonder, and read stories of the voyages of Marco Polo, Christopher Columbus, and other great explorers, always finding myself looking for a new place to see and a new adventure to find, trying to capture the feeling of discovering a new place, of finding a new horizon to uncover. I see this mission as a direct continuation of those same early pioneers, embodying the true meaning of adventure."
>
> —Yair Maimon, Mars One applicant

But what will happen when interest in the show wanes? Viewership of any program can be cyclical; ratings rise and fall as novelty peaks and fatigue sets in over time. It seems inevitable that Earth audiences will eventually seek entertainment elsewhere, whether it's because the Martian broadcasts have become more mundane or other shows or options have become more enticing.

When viewers start tuning out, will the settlers feel increasingly frustrated, isolated, and depressed? It's possible . . . but, hopefully, the settlers will counteract these feelings by focusing more attention on other leisure-time activities, such as recording their experiences in ways that aren't quite as ratings-sensitive as reality-TV shows.

The Blogger Chronicles

Like the HI-SEAS participants, the Mars One settlers will no doubt blog often about their Martian experience. Blogging is an ideal medium because it doesn't rely on real-time interaction or streaming to convey content. Bloggers on Mars will be able to file their posts and walk away from the keyboard, just like the bloggers on Earth who write and post on extended, irregular schedules.

Blogging will provide a vital, open record of settlers' experiences for readers back home, even as it allows the settlers to vent and ponder the parameters of their new lives. If anyone prefers to keep their thoughts more private, they'll be able to keep personal journals and notes instead.

Such documentation, besides its therapeutic effects for the writers themselves, will capture the many details of the Martian adventure for posterity. Notable "firsts" are sure to be recorded, from the sublime to the ridiculous—everything from the first sunset watched by humans on another world to the first joke told on Mars.

Other firsts might not make it into the public record, though. At some point, settlers are bound to turn to "adult" pursuits, like alcohol consumption and sex, to relieve boredom and stress. Excessive indulgence in such activities has been seen before, in other isolated settings. Personnel posted at the Amundsen-Scott South Pole Station in Antarctica, for example, have been known to drink heavily on a regular basis just to endure the long winter months.

Such activities might not be recorded in blogs or journals, unless they got out of hand in some way and posed a danger to the settlement. In that case, settlers might have to spend their free time on achieving another first that will be critical to the long-term success of the mission: the development and enforcement of a Martian legal system, complete with punishments and deterrents for negative behavior.

The creation and implementation of a legal system on Mars is sure to be a historic event, but other, recreation-related firsts will be just as compelling in their own way. For wherever humans go, they bring their foibles and laws along with them . . . and their games, as well.

There'll be a first golf ball teed off, of course, in the tradition of Alan Shepard on the moon . . . and other sports are sure to follow. Even wearing Mars Suits to protect themselves against the hostile environment, the settlers will find that performing amazing athletic feats will be irresistibly easy, with gravity at a mere 38 percent of the gravity they left behind on Earth.

The Ultimate Sports Fantasy

Though the settlers won't have brought much extra cargo with them on the flight from Earth, you can bet someone will have stowed baseballs and a bat on the spacecraft. How could the Mars One team resist a chance to hit the ultimate home run in the low Martian gravity?

Though the mass of a baseball on Mars will be the same as on Earth, the ball's weight will be about one-third of its Earth weight. The reduced weight will let the ball soar much farther after being hit than it would on Earth—and the thinner atmosphere will play a role, too. At six hundred pascals, the atmospheric pressure on Mars is only about 0.6 percent of the mean sea-level pressure on Earth. This lower pressure will reduce the drag on a baseball in motion, thus extending its flight even farther.

Because the same factors will apply to other hurled objects, can there be any doubt that the settlers will bring along other sports equipment as well? Imagine the loft and distance of a football pass on Mars, or a struck tennis ball. You just *know* the settlers will use some of their precious leisure time to set some serious sports records.

Target shooting might be another sport of interest on Mars, as the lower gravity and thinner atmosphere will change the dynamics of projectiles in flight. How much farther will a bullet travel than if it were fired from the same gun on Earth? Eight to ten times farther, according to some estimates, but it will be interesting to see the actual results as they happen.

And viewers back home will surely see it all. Segments on the farthest home run, football pass, and target shoot of all time will be a perfect fit for the Mars One reality show. Folks on Earth will eat it up, and the settlers will have fun playing while demonstrating the properties of their new environment.

Though other demonstrations of those properties will also take up some of their free time, "fun science" is bound to be part of the schedule.

Mr. Wizards of Mars

For years, astronauts on the International Space Station and space shuttle flights have conducted ancillary experiments conceived by students—demonstrations that illustrate the unique behaviors of various materials and organisms in an off-Earth environment. These experiments are vetted and arranged by the Student Spaceflight Experiments Program (SSEP), sponsored by the National Center for Earth and Space Science Education (NCESSE), the Arthur C. Clarke Institute for Space Education, and NanoRacks. The initiative is designed "to inspire and engage America's next generation of scientists and engineers."

The same kind of experiments will no doubt be conducted by the Mars One pioneers in their spare time. Based on student experimental designs, the settlers might explore crystal growth or seed germination under the unique gravitational conditions on Mars. They might study the behavior of microorganisms to see if any are equipped to survive in the thin Martian atmosphere. They might also demonstrate the impact of Martian conditions—low gravity, low atmospheric pressure, intense solar radiation, frigid temperatures—on the biology of plant and animal cells brought from Earth for study and experimentation.

"I want to do something that matters; something that both myself and those that I care about can be proud of. I want them to be proud to say they had inspired me."

—Mars One applicant

But the spare-time lab work won't end there. Like shuttle and space station astronauts before them, the settlers might dabble in less structured experiments. In the spirit of the children's TV science show host Mr. Wizard, the settlers might get a kick out of illustrating how the properties of Mars affect the properties of different substances, all recorded on video for the benefit of viewers back home. They might demonstrate, for example, how certain chemical reactions play out differently on Mars than on Earth. They might show how a compass behaves on Mars, which doesn't have a magnetic field, or experiment with the way sound carries in the sparse atmosphere. These "fun science" exercises would provide important information as well as a welcome distraction . . . and a way to engage future Mars settlers watching from homes and classrooms on Earth.

But another effective way to engage an Earth audience and lay the groundwork for future settlement might come from a very different type of leisure-time activity—one more concerned with *subjective* results than *objective* ones.

Interplanetary Creativity

If there's one thing we can bet on, it's this: It won't take long for the Martian settlers to start capturing their experiences through creative expression.

As surely as the human heart is touched by joy and hardship, victory and tragedy, beauty and ugliness, it seeks to translate these things into art. Whether the artist means to interpret the truth of an experience, come to terms with its emotional impact, or just capture its reality as completely as possible for future reference, he or she is ultimately compelled to use the stuff of daily life as raw material for new creations.

With lives like the ones the first Martian settlers will have, the raw material will be very rich indeed. As the first human beings to inhabit another world, how can they *not* be inspired to express their unique experiences in creative ways?

Depending on their individual talents, the settlers will most likely be moved to write, draw, paint, or compose music. Such creative pursuits will help to make the isolation and confinement bearable while simultaneously making the Mars One experience resonate for the people back home.

The works that come out of this creativity might have a better chance of capturing the imagination of humanity than all the dry facts and observations that will arise from inhabitation and study of the Red Planet.

A well-written poem can inspire in ways no technical report ever can. A memorable story can set hopes and dreams in motion in ways beyond the reach of any stodgy scientific treatise. The right song can motivate millions and last forever.

These and other art forms hold the power to win the hearts of an Earth audience and generate support for future missions to Mars. These art forms have the potential to plant seeds in the minds of the right people, sparking them to someday step forward and play key roles in the Mars One program.

Art speaks to the soul in ways that facts and figures never can, and the unique art that swirls to life on Mars will speak with a voice that has never been heard in the same way before. Just imagine what the settlers will be able to create with a canvas as big as a world.

The Face on Mars Winks

Works of art developed on Mars will reflect the uniquely Martian experience of settlers and conditions that exist only in that one place in the entire solar system. Early works will likely incorporate the landscape of the Red Planet as filtered through the human imagination—the way the world looks and feels from inside protective habitats and Mars Suits. Prose, poetry, paintings, and music will interpret what exists beyond the shells in which the settlers live, overlaying human consciousness upon an environment that is always removed from direct human contact.

But as life on Mars becomes more routine and leisure time more plentiful, the scale of artistic ambitions will no doubt increase. Though still required to wear Mars Suits to survive outside, settlers will begin to find ways to interact more directly with the environment, taking fuller advantage of its vastness and physical properties to make their creations more unique.

They might carve blocks of Martian rock into huge sculptures or stack them into monolithic Stonehenge-like monuments. Enormous figures or abstract designs could be scrawled across tracts of land in the style of the Nazca Lines or crop circles of Earth. (How cool would it be for humans to draw otherworldly crop circles on the surface of another planet?)

What about using giant sheets of fabric to cover land formations? Artists Christo and Jeanne-Claude have done similar work on Earth, draping fabrics of varying types and colors over buildings and landforms. In 1983, for example, Christo and Jeanne-Claude surrounded eleven islands in Biscayne Bay in Miami, Florida, with pink fabric. A more recent project involved covering a 10.8-kilometer-long section of the Arkansas River with suspended fabric panels. Perhaps someday, Mars settlers will do something similar. Imagine blue streamers strung for kilometers over the Martian surface, simulating from a distance the water-filled canals that early astronomers once thought they glimpsed there.

"At the end of the day, humankind is a miracle for me to behold and I believe in us. I am proud to be human and I am proud to know Mars One is a legacy we will all share together as a human race. I would be honored and humbled to be chosen as one of the first to go and inhabit a new future for mankind."

—Mars One applicant

What about filmmaking? The vast, barren landscapes will make the perfect locations for shooting movies. Filmmakers, and every breed of creative type, will have an entire empty planet to serve as the backdrop for—or star of—their projects.

And imagine the incredible dance productions that could be staged in 38 percent of Earth's gravity, with red skies, plains, and mountains all around. New and sleeker Mars Suits would have to be invented, but dancers on the Red Planet might someday put Cirque du Soleil to shame.

Musicians could do groundbreaking work in the unique Martian environment, too. What might a classical composition sound like when performed in the thin, exotic mix of gases in the Martian atmosphere? Would musical instruments have to be modified to generate any kind of tuneful sound outside the sealed habitats? Imagine entire symphonies attuned to the unique properties of the atmosphere on Mars, played by settler orchestras on modified instruments, sounding like nothing ever heard before by human ears.

Perhaps such music could complement dazzling light shows projected onto Martian landmarks. With the right equipment and power sources, artistic settlers might be inspired to stage such displays, illuminating natural wonders with different colors and patterns of radiant laser light. Imagine such a show projected onto the slope of Olympus Mons in some distant Martian future:

Dancing green lasers bring to life an enormous figure on the side of the massive volcanic mountain—Tars Tarkas, the original four-armed, green-skinned warrior of Edgar Rice Burroughs' Barsoom stories. The lasers flash and swoop, changing the image they project from Tars Tarkas to a tripod killing machine, one of the Martian invaders of Earth in H. G. Wells' War of the Worlds.

Then the lasers change once more, tracing the image of a bespectacled man's face. It's a familiar face, a quintessentially Martian *face, though the man it belonged to never once set foot on the Red Planet. Perhaps as much as anyone, he brought Mars home to generations of Earthlings, kept the soul of it alive in their hearts and reinforced the vision of humanity alive and well under alien red skies. He was a creator, and the modern-day Martians pay tribute to him with their own creative magic.*

The lights flash and the image moves. The face of the author of The Martian Chronicles, *Ray Bradbury, painted with lasers on the slope of the great mountain, winks at anyone who's watching from Mars or Earth or anywhere else among the glittering stars, sprinkled like diamonds in the night sky above.*

Robert T. Jeschonek*'s love of Mars started in childhood with Edgar Rice Burroughs and Ray Bradbury, continued with Kim Stanley Robinson, and is stronger than ever thanks to the Curiosity rover and the brave new Mars One program. His essays have been featured in Smart Pop's* Fringe Science, In the Hunt: Unauthorized Essays on Supernatural, *and* House Unauthorized. *His science fiction and fantasy stories have appeared around the world in publications including* Galaxy's Edge, Space and Time, *and* Escape Pod. *His novels have won the International Book Award, the Forward National Literature Award, and the Scribe Award. Visit him online at www.robertjeschonek.com. You can also find him on Facebook and follow him on Twitter as @TheFictioneer.*

THE MARS ONE MISSION TIMELINE

2011	Mars One Founded	Bas Lansdorp and Arno Wielders laid the foundation to begin the Mars One mission plan. The first steps included holding discussion meetings with potential aerospace component suppliers in the United States, Canada, Italy, and the United Kingdom. The mission architecture, budgets, and timelines were then solidified after receiving feedback from the supplier engineers and business developers. This resulted in a baseline design for an achievable mission of permanent human settlement on Mars.
2013	Crew Selection Commences	The Astronaut Selection Program (ASP) was launched at press conferences in New York and Shanghai. The selection program required an online application and proceeded with video applications and personal interviews. The subsequent selection rounds will consist of group challenges and simulations. At program's end, scheduled for winter of 2016, six teams of four individuals will have been chosen for training. New ASPs will begin every year to replenish the training pool regularly.

2017	Crew Training Commences	Groups selected from the first batch of applicants will train together until the launch in 2026. The groups' ability to deal with prolonged periods of time while sequestered in a remote location will be the most important part of their training. Thus, they will learn how to repair components of the habitat and rover, train in medical procedures, and learn to grow food in the habitat. Every group spends several months of each training year in the analog outpost to prepare for its mission to Mars.
Unmanned Missions 2020–2026		
2020	Demo and Comsat Mission	A demonstration mission to Mars will be launched in 2020. This mission will provide proof of concepts for some of the technologies that are crucial for a human mission to Mars. A communication satellite will also be launched, which will be placed into stationary Mars orbit. This satellite will enable communications between Earth and Mars on a 24/7 basis, except when the sun is between the two planets, and can also relay images, videos, and other data from the Mars surface.
2022	Rover Mission Launched	One rover and one trailer will be sent to Mars. The rover can use the trailer to transport the landers to the outpost location. On Mars, the rover drives around the chosen region to find the best location for the settlement. Ideally, the settlement can be located far enough north for the soil to contain enough water, close enough to the equator to produce maximum solar power, and on flat enough terrain to facilitate the construction of the settlement.

		When the settlement location is determined, the rover will prepare the surface for the arrival of the cargo missions. It will also clear large areas for the placement of solar panels. A second communications satellite will be launched into orbit around the sun. It will follow the same orbit as Earth does but trail 60 degrees behind it in the L5 Lagrangian point of the Sun-Earth system. Together with the first Comsat in Mars orbit, they will enable 24/7 communication with Mars, even when the sun is between the two planets.
2024	Cargo Mission Launched	A second rover, two living units, two life support units, and a supply unit will be sent to Mars. In 2025, all units will land on Mars using a rover signal as a beacon.
2025	Outpost Operational	The rover will pick up the first life support unit using the trailer, place the life support unit in the correct location, and deploy the thin-film solar panel of the life support unit. The rover will then be able to connect to the life support unit to recharge its batteries much faster than using only its own panels, which will allow it to work much more efficiently. The rover will pick up all the other cargo units and deploy the thin-film solar panel of the second life support unit and the inflatable sections of the living units. The life support unit will be connected to the living units by a hose that can transport water, air, and electricity. Once these are connected, the Environmental Control and Life Support System (ECLSS) will be activated. The rover will feed Martian soil into the ECLSS, and the water will be extracted from this soil by evaporating its subsurface ice particles

		in an oven. The evaporated water is condensed back to its liquid state and stored, and part of the water is used for producing oxygen. The nitrogen and argon filtered from the Martian atmosphere will make up the other components of the breathable air inside the habitat. Before the first crew begins its journey, the ECLSS will have produced a breathable atmosphere of 0.7 barometric pressure, 3,000 liters of water, and 240 kilograms of oxygen, which will be stored for later use. The rover will also deposit Martian soil on top of the inflatable sections of the habitat to shield it from radiation.
Manned Missions 2026–First Settlement 2027		
2026	Crew One Departs	The components of the Mars Transit Vehicle (MTV) will be launched into Earth orbit after receiving the green light on the status of the systems on Mars. First, a transit habitat and a Mars lander with an assembly crew aboard will be launched into Earth orbit. The assembly crew will dock the Mars lander to the transit habitat. About thirty days later, two propellant stages, the boosters that will "kick" the transit vehicle from low Earth orbit to Mars transfer orbit, will be launched and connected. Second, the first fully trained Mars crew will be launched into the same Earth orbit. Once in orbit, this crew will switch places with the assembly crew, which will descend back to Earth. After a final check of systems on Mars and on the transit vehicle, engines of the propellant stages will be fired and the MTV will be launched into a Mars-transit trajectory. This is the point of no return for the Mars crew.

		The cargo for the second crew will be launched toward Mars in the same month as the launch of the first Mars settlers.
2027	First People Land on Mars	Approximately twenty-four hours before landing, the crew will move from the transit habitat into the landing module, bringing some of the supplies from the transit habitat. The landing module will then detach from the transit habitat. (The lander is exactly the same as those used for previous unmanned missions. This will ensure that the human crew lands in a system that has been tested eight times before.) Upon landing, the crew members will take up to forty-eight hours to reacclimate to gravity after spending six to eight months in space. They will leave the lander in their Mars Suits and be picked up by the rover. Next, the astronauts will enter the settlement through an air lock and spend the next few days in one of the living units, recovering and settling into their new environment. The settlers will deploy the rest of the solar panels after their acclimatization period. They will also install the hallways between the landers and set up food production units. The cargo for the second crew will land within a few weeks after the first crew has landed. This cargo will also be picked up and installed. When the first crew lands, it will find the established habitat with good redundancy as, by this time, it will include two living units, each large

		enough to house the crew of four, and two life support units that are capable of providing enough water, power, and breathable air for the entire crew. When the hardware for the second crew is incorporated into the settlement, it will feature four living units and four life support units, which are enough to sustain a crew of sixteen astronauts.
2028	Crew Two Departs	The second crew will depart from Earth. The cargo for the third crew is launched with the second crew. The second crew will land on Mars in 2029. The hardware for the third crew will land a few weeks later and be added to the settlement. This process will continue as additional crews land every two years.

THE MARS ONE SELECTION PROCESS

Round One (April 2013–December 2013)

When the application program for the Mars One mission to Mars was announced in 2013, 202,586 applicants filled out the online registration, including email address, country of residence, and date of birth. Each individual also agreed to the legal terms and conditions of the application program.

The applicants had to overcome a number of hurdles during the process to show their commitment to embark on the one-way mission to Mars. The first was to pay a small administration fee, which was calculated based on the GDP (gross domestic product) of the applicant's country of residence.

After submitting the payment, applicants received access to the full application form, which included various steps divided into public and private components. The public component included submitting a profile image, basic personal information, and a video in which the applicant answered various questions that addressed why he or she was applying to the mission, how the applicant described his or her sense of humor, and why the applicant felt he or she would make a good candidate for settling Mars. The private component included more in-depth personal information, as well as an essay on the applicant's motivation for applying. The private component also required answers to open-ended questions about how the applicant had handled difficult and stressful situations in the past. Both components helped the members of the Mars One Selection Committee (MOSC) determine if the applicants understood what they were applying for and were sincere about settling on Mars.

Completing the entire application was necessary to move forward in the selection process. The total number of fully and correctly completed applications was 4,227. After a complete review, the MOSC chose 1,058 applicants from around the world to move on to Round Two.

Round Two (January 2014–February 2015)

In Round Two, the 1,058 remaining candidates were required to take a medical exam conducted by their own physician and then submit the medical statement, signed and endorsed by the physician, to the MOSC. This medical exam is very similar to those required by NASA and the European Space Agency. Among other things, the candidates were examined for blood and urine parameters; 20/20 visual acuity in both eyes, either uncorrected or corrected with lenses or contact lenses; good general health with an age- and gender-adequate fitness level; a BMI of less than thirty; blood pressure not to exceed 140/90, measured in a sitting position; lack of drug dependency; and the ability to execute full range of motion and function in all joints. (One of the benefits of the medical exam was that some people were alerted to a medical condition that required treatment.)

Of the 1,058 candidates that entered Round Two, 660 of them successfully completed the medical exam and obtained physician approval and were invited for an interview. This brief (15 minutes, on average), carefully structured video interview was designed to screen out those who were less likely to fulfil Mars One's other requirements. This interview included open-ended questions about Mars One's overall mission and candidates' teamwork experience and family support, in addition to knowledge of questions drawn from material that candidates were required to study from the Mars One website. Applicants received one month's notice about which pages from the Mars One website would be relevant in the interview. Each candidate then had a two-month period to select the day and time for his or her interview.

The knowledge questions helped determine not only whether a candidate was serious about the project, but also if the person was a strong learner—could he or she retrieve, retain, and apply knowledge (essential for Mars settlers). The material chosen also helped ensure that candidates really understood some of the dangers and risks involved in a mission to Mars. They were asked to memorize facts, such as how much radiation they would be exposed to, how much shielding would be needed, how long the reserve water and oxygen would last, and which noncritical systems could be shut down or restricted to save energy.

The open-ended interview questions also helped determine the likelihood that a candidate would be a good team player. For example, questions such as "Tell me a story from your life that shows what unique, valuable qualities you brought to a team," or "After three years on Mars, were a return flight possible, would you take the trip?" were designed to extract whether or not a candidate would really be likely to put the team ahead of him- or herself. In an effort to further ascertain sincerity and seriousness about settling on Mars in service to humanity, candidates were also asked to answer the following: "Tell me about the day when you decided to apply to settle on Mars, with no return to Earth, and why you made that decision."

From 660 Round Two candidates, one hundred of them were selected in February 2015 to move forward to Round Three.

Round Three: A Work in Progress

From this milestone onward, selection and training activities will be filmed for viewing by audiences across the globe. Filming is planned to commence sometime in 2016.

During Round Three, the remaining one hundred candidates will be trimmed down to twenty-four. The final twenty-four will receive a formal full-time employment offer to train for the mission to Mars. Those who accept will begin a ten-year training program as six teams of four. A lot can happen in a decade; some participants might not perform as hoped and others might quit the program, so the MOSC will continue screening applicants throughout this training period. (Remember, a participant is only guaranteed a trip to Mars once they are on a rocket that is leaving Earth.) Every two to three years, new selection rounds will be initiated in order to resupply the pool of Mars settlers.

The Mars 100: Group Challenge

The one hundred candidates who were selected for Round Three will convene at a location chosen by the MOSC. To allow them to get better acquainted with one another, the candidates will first self-select into six to ten teams of ten to fifteen members. They will be instructed to group themselves into teams with the people they believe they can work well with, but each team needs to be as diverse as possible with regard to age, nationality, and ethnicity and composed of an equal number of men and women.

The MOSC will set up group-dynamic challenges and provide study materials related to the challenges, which will allow it to observe how the candidates work in a group setting. How did the candidates organize themselves into teams? How well did they solve problems as a team? How did each candidate handle the conflicts that inevitably emerge when facing a challenge together? Personality characteristics should rise to the surface during this process. The MOSC members will watch all of this and debrief the candidates afterward. Every day, ten to twenty candidates will be eliminated.

This process will play out throughout five days, and from its observations, the MOSC will be able to whittle down the potential candidates from one hundred to forty.

The Mars 40: Isolation Challenge

The forty candidates who pass the group-dynamics challenge will then begin the nine-day isolation portion of the screening process. But first they will have to fill out an extensive questionnaire about their preferences—personal and living.

During a long voyage together and afterward, living in permanent settlement, a small group of people can't hide from or avoid one another. This means the little things can matter quite a bit—dirty socks on the floor can bother people, for example, or dirty dishes left in the sink or pungent body odors. Though problematic team members should have been eliminated by this stage, certain team members are likely to click together better than others. Finding the most compatible combinations of four will be essential to the success of the Mars One mission.

In the isolation unit, the forty candidates will be tested on the study materials provided to them beforehand that related to the final set of challenges. The material will cover tests on broken systems, emergency situations, and team decision-making, as well as discussions about current political topics. Candidates will also be tested on self-grown food preparation and tasting, because experiential learning is an essential part of what the final twenty-four candidates will be engaging in throughout the next ten years. The results of the isolation challenge will pare the forty candidates down to thirty who will then undergo the Mars Settler Suitability Interview (MSSI).

The Mars 30: The Mars Settler Suitability Interview

The Mars Settler Suitability Interview measures the suitability of a candidate for long-duration space missions and Mars settlement. The MSSI includes, but is not limited to, the following topics: teamwork and group living skills, motivation, family issues, performance under stressful and unique working conditions, judgement, and decision-making. The interview lasts four hours and will be video recorded for analysis by the MOSC members. From these thirty candidates, the final twenty-four will be selected and offered full-time employment to begin mission training.

The Final 24: Mission Training

Training for the final twenty-four crew members will comprise Phase One, technical, and Phase Two, personal and group.

Phase One: Technical Training

Phase One will include the training of all crew members for proficiency in the use and repair of all equipment to the extent that they can identify and solve technical problems. In addition, all crew members will receive extensive medical training to treat chronic, acute, and critical health problems. At least two crew members will study the geology of Mars and building construction, and two crew members will gain expertise in exobiology, which concerns the effects of extraterrestrial environments on living things. Furthermore, all twenty-four crew members will study law and political science.

Phase Two: Personal and Group Training

Phase Two training will mainly involve simulation missions, during which the crew members will take part in fully immersive exercises that prepare them for the real mission to Mars. The simulated environment will closely match that of planet Mars. The astronaut-teams-in-training will participate in these simulations for a few months each year.

Selection Committee Member Requirements

Choosing Mars One participants requires a high level of expertise. Since they come from all over the globe and will undertake the long journey to Mars to live there for the rest of their lives, the members of the Mars One Selection Committee must possess an understanding of different cultures, as well as have many years of experience working with astronauts, extreme environments, and—of utmost importance—isolated habitats.

The following mandatory minimum requirements shall be met to be a member of the MOSC:

- Degrees: MD, PhD, or equivalent education
- Published as lead author of peer-reviewed content about the field of human spaceflight (long-duration spaceflight/group dynamics)
- Fifteen or more years of professional experience working in the field of human spaceflight (group dynamics/long-duration spaceflight/medicine/psychology/psychophysiology)
- Fifteen or more years of professional experience working with astronauts or astronaut candidates who come from different continents
- Experience working with active astronauts from different space agencies
- Experience working for different space agencies
- Experience conducting experiments with isolation chambers or equivalents as a principal investigator

The additional preferred requirements to be met are as follows:

- Experience working actively on the human spaceflight requirements for the International Space Station or equivalent
- An award from an aerospace association

The Current Selection Committee Members

Norbert Kraft, MD, Chief Medical Officer (USA)
Professor Raye Kass, PhD, Advisor (Canada)
James R. Kass, PhD, Advisor (Netherlands)

ENDNOTES

TECHNICAL AND MEDICAL SKILLS, HEALTH AND FITNESS

"Human Health and Performance," Jamie R. Guined

1. NASA, *Human Exploration of Mars, Design Reference Architecture 5.0, Addendum* (SP-2009-566-ADD: 2009), accessed October 18, 2014, http://www.nasa.gov/pdf/373667main_NASA-SP-2009-566-ADD.pdf.
2. NASA, *Human Research Program Requirements Document, Revision F* (HRP-47052: 2013), accessed October 18, 2014, http://www.nasa.gov/pdf/579466main_Human_Research_Program_Requirements_DocumentRevF.pdf.
3. Richard S. Johnston and Lawrence F. Dietlein, eds., *Biomedical Results from Skylab*, NASA SP-377 (Washington, DC: US Government Printing Office, 1977), 18.
4. John R. Allen, "Sensory-Motor Issues Related to Space Flight," PowerPoint presentation, from NASA Headquarters, *Space Transportation System (STS) 130 Endeavor Crewmembers*, March 2010, http://myavaa.org/documents/conferences/AVAA-March-2010-Conference/PDF-Presentations/Allen%20Wed%200800%20JDVAC%202010.pdf.

CULTURE, COHESION, AND COMPATIBILITY

1. Norbert O. Kraft, Terence J. Lyons, and Heidi Binder, "Intercultural Crew Issues in Long-Duration Spaceflight," *Aviation, Space, and Environmental Medicine* 74, no. 5 (2003): 575–78.
2. Ibid.

"A World Waiting to Be Born," Raye Kass

1. "Shackleton's Ad—Men Wanted for Hazerdous Journey," Discerning History, last modified on May 15, 2013, accessed May 5, 2015, http://discerninghistory.com/2013/05/shackletons-ad-men-wanted-for-hazerdous-journey.
2. Caroline Graham, "'For 14 days we were all in pitch darkness. There was no night and no day. We begged God to help us': The amazing first interview with one of the trapped Chilean miners," *The Daily Mail*, last modified October 17, 2010, http://www.dailymail.co.uk/news/article-1321230/Chilean-miners-World-exclusive-interview-Mario-Sepulveda.
3. Roland Huntford, *Shackleton* (London: Hodder & Stoughton, 1985), 461.
4. Sir Ernest Shackleton, *South: The Story of Shackleton's Last Expedition, 1914–1917* (London: Century Publishing, 1983), 81–82.
5. "Chile Miners Reveal Dark Times Underground," *The Australian News*, last modified October 15, 2010, http://www.theaustralian.com.au/news/world/chile-miners-reveal-dark-times-underground/story-e6frg6so-1225939146087.
6. Ibid.
7. Rory Carroll and Jonathan Franklin, "Chile Miners: Rescued Foreman Luis Urzúa's First Interview," *The Guardian*, last modified October 15, 2010, http://www.theguardian.com/world/2010/oct/14/chile-miner-luis-urzua-interview.

8. Caroline Graham, "'For 14 days we were all in pitch darkness'."

9. Rory Carroll and Jonathan Franklin, "Foreman Luis Urzúa's First Interview."

10. Valentin Vital Evich Lebedev, *Diary of a Cosmonaut: 211 Days in Space* (College Station, TX: Phytoresource Research Inc, Information Service, 1988).

11. Personal communication, Prof. I. Samaltedinov, Division of Psychology, Training Department, Star City, Moscow (1994).

12. Jack Stuster, *Bold Endeavours: Lessons from Polar and Space Exploration* (Annapolis, MD: Naval Institute Press, 1996), 165.

13. Sven Grahn, "Cosmonaut Vladimir Vazyutin—A Different Kind of Space Hero," accessed September 12, 2015, http://www.svengrahn.pp.se/trackind/scramble/Vazyutin.htm.

14. Martha Freeman, *Challenges of Human Space Exploration* (New York: Springer Praxis, 2000), 7–9.

15. Raye Kass, *Theories of Small Group Development*, 4th ed. (Montreal: Concordia University, Centre for Human Relations and Community Studies, 2008).

16. Daniel Goleman, *Working with Emotional Intelligence* (New York: Random House, 1998), 94.

17. Ibid., 4.

18. Raye Kass and James Kass, "Team Work During Long-Term Isolation: SFINCSS Experiment GP 006," in *Simulation of Extended Isolation: Advances and Problems*, ed. Victor M. Baranov (Moscow: Firm SLOVO, 2001), 124–47.

19. N. Kraft, T. Lyons, H. Binder, "Intercultural Crew Issues in Long-Duration Spaceflight," *Aviation, Space, and Environmental Medicine* 74, no. 5 (2003): 575–78.

20. Ibid.

21. Larry Bell, "A Conversation with Buzz Aldrin Regarding His Vision for Space Exploration," *Forbes*, last modified July 9,

2013, http://www.forbes.com/sites/larrybell/2013/07/09/a-conversation-with-buzz-aldrin-regarding-his-vision-for-space-exploration.

"Culture and Communication," Andy Tamas

1. K. L. Van der Zee and J. P. Van Oudenhoven. (2000). "The Multicultural Personality Questionnaire: A Multidimensional Instrument of Multicultural Effectiveness." *European Journal of Personality*, 14 (4), 291–309; K. L. Van der Zee and J. P. Van Oudenhoven. (2001). "The Multicultural Personality Questionnaire: Reliability and Validity of Self- and Other Ratings of Multicultural Effectiveness." *Journal of Research in Personality*, 35 (3), 278–288.
2. Adapted from Joyce L. Hocker and William W. Wilmot, "Conflict Styles," in *Interpersonal Conflict* (Iowa: W. C. Brown Company, 1978), 23–45.

"Age and Aging on Mars"

1. Quotes here are not attributed directly to candidates, but the candidates who contributed thoughts are listed in the book's acknowledgments. Quotations have been modified to aid readability.

"Men Are from Mars, Women Are from Mars," Ronit Kark

1. Mark, S., G. B. Scott, D. B. Donoviel, L. B. Leveton, E. Mahoney, J. B. Charles, and B. Siegel. 2014. "The Impact of Sex and Gender on Adaptation to Space: Executive Summary." *Journal of Women's Health* 23, no. 11 (2014): 941–47.
2. Ibid.
3. Hiler, K. 2013. "NASA's New Class of Astronauts Gives Parity to Men and Women," *New York Times*, June 18.
4. Phelan, E. J., C. A. Moss-Racusin, and L. A. Rudman. 2008. "Competent Yet Out in the Cold: Shifting Criteria for Hiring

Reflect Backlash Toward Agentic Women." *Psychology of Women Quarterly* 32: 406–13.

5. Lyness, S. K., and M. E. Heilman. 2006. "When Fit Is Fundamental: Performance Evaluations and Promotions of Upper-Level Female and Male Managers." *Journal of Applied Psychology* 91: 777–85.

6. Boldry, J., W. Wood, and D. A. Kashy. 2001. "Gender Stereotypes and the Evaluation of Men and Women in Military Training." *Journal of Social Issues* 57: 689–705; Heilman, E. M., and A. S. Wallen. 2010. "Wimpy and Undeserving of Respect: Penalties for Men's Gender Inconsistent Success." *Journal of Experimental Social Psychology* 46: 664–67; Wiesel, V., and R. Kark. 2014. "How a 'Tailor Made' Manager Suit, Made for Male Candidates, Fits Female Candidates?" Not published.

7. Spencer, H. October 2009. "Why NASA Barred Women Astronauts." *Short Sharp Science*, https://www.newscientist.com/blogs/shortsharpscience/2009/10/why-nasa-barred-women-astronau.html.

8. Eagly, A. H., and L. L. Carli. 2007. *Through the Labyrinth: The Truth About How Women Become Leaders*. Boston: Harvard Business School Press; Heilman, E. M., A. S. Wallen, D. Fuchs, and M. M. Tamkins. 2004."Penalties for Success: Reaction to Women Who Succeed at Male Gender-Typed Tasks. *Journal of Applied Psychology* 89: 416–27.

9. Cuddy, A. J. C., S. T. Fiske, and P. Glick. 2004. "When Professionals Become Mothers, Warmth Doesn't Cut the Ice." *Journal of Social Issues* 60: 701–18; Eagly, A. H., and S. J. Karau. 1991. "Gender and the Emergence of Leaders: A Meta-analysis." *Journal of Personality and Social Psychology* 60: 685–710; Kark, R., and A. H. Eagly. 2010. "Gender and Leadership: Negotiating the Labyrinth." In *Handbook of Gender Research in Psychology*, edited by J. C. Chrisler and D. R. McCreary, 443–68. New York: Springer.

10. Glick, P., and S. T. Fiske. 2001. "An Ambivalent Alliance: Hostile and Benevolent Sexism as Complementary Justifications for Gender Inequality." *American Psychologist* 56: 109–18; "Hostile and Benevolent Sexism: Measuring Ambivalent Sexist Attitudes Toward Women." *Psychology of Women Quarterly* 21: 119–35.

11. Diekman, B. A., A. E. Eagly, and A. M. Johnston. 2010. "Social Structure." In *The Sage Handbook of Prejudice, Stereotyping and Discrimination,* edited by J. F. Dovidio et al., 209–24. London: Sage Publications.

12. Mark et al. "The Impact of Sex and Gender On Adaptation to Space."

13. Budig, M. J. 2002. "Male Advantage and the Gender Composition of Jobs: Who Rides the Glass Escalator?" *Social Problems* 49: 258–77; Risman, B. J., J. Lorber, and J. H. Sherwood. 2012. "Toward a World Beyond Gender: A Utopian Vision," prepared for the August 2012 American Sociology Society Meeting: 17–20.

14. Miller, W., B. Kerr, M. Reid. 1999. "A National Study of Gender-Based Occupational Segregation in Municipal Bureaucracies: Persistence of Glass Walls?" *Public Administration Review*: 218–30.

15. Ryan, M. K., and S. A. Haslam. 2005. "The Glass Cliff: Evidence That Women Are Over-represented in Precarious Leadership Positions. *British Journal of Management* 16 (2): 81–90.

16. Fletcher, J. K. 2004. "The Paradox of Postheroic Leadership: An Essay on Gender, Power, and Transformational Change." *Leadership Quarterly* 15: 647–61.

17. Woolley, A., and T. Malone. 2011. "What Makes a Team Smarter? More Women." *Harvard Business Review* 89 (6): 32–3; Woolley, A. W., C. F. Chabris, A. Pentland, N. Hashmi, and T. W. Malone. 2010. "Evidence for a Collective Intelligence Factor in the Performance of Human Groups." *Science* 330 (6004): 686–88.

18. Hong, L., and S. E. Page. 2004. "Groups of Diverse Problem Solvers Can Outperform Groups of High-Ability Problem Solvers." *Proceedings of the National Academy of Sciences* 101 (46): 16385–89.

19. van Knippenberg, D., and M. C. Schippers. 2007. "Work Group Diversity." *Annual Review of Psychology* 58: 515–41; Eagly. "Gender and Leadership."

20. Joshi, A. 2014. "By Whom and When Is Women's Expertise Recognized? The Interactive Effects of Gender and Education in Science and Engineering Teams." Abstract. *Administrative Science Quarterly*.

21. Hoogendoorn, S., H. Oosterbeek, and M. Van Praag. 2011. "The Impact of Gender Diversity on the Performance of Business Teams: Evidence from a Field Experiment." Tinbergen Institute Discussion Paper, no. 11-074/3.

22. Adams, R. B., and D. Ferreira. 2009. "Women in the Boardroom and Their Impact on Governance and Performance." *Journal of Financial Economics* 94 (2): 291–309.

23. Kark, R., R. Waismel-Manor, and B. Shamir. 2012. "Does Valuing Androgyny and Femininity Lead to a Female Advantage? The Relationship Between Gender-Role, Transformational Leadership and Identification." *The Leadership Quarterly* 23 (3): 620–40.

24. Ely, R. J., and Meyerson, D. E. "Unmasking Manly Men." *Harvard Business Review* 86 (July/August 2008): 20; "An Organizational Approach to Undoing Gender: The Unlikely Case of Offshore Oil Platforms." *Research in Organizational Behavior* 30: 3–34.

25. Ibid.

26. Dovidio, J. F., C. E. Brown, K. Heltman, S. L. Ellyson, and C. F. Keating. 1988. "Power Displays Between Women and Men in Discussions of Gender-Linked Topics: A Multichannel Study." *Journal of Personality and Social Psychology* 55: 580–87.

27. Eagly, A. H., and S. J. Karau. 1991. "Gender and the Emergence of Leaders: A Meta-analysis." *Journal of Personality and Social Psychology* 60: 685–710.

28. Bowles, H. R., L. Babcock, and L. Lai. 2007. "Social Incentives for Gender Differences in the Propensity to Initiate Negotiations: Sometimes It Does Hurt to Ask." *Organizational Behavior and Human Decision Processes* 103 (1): 84–103.

29. Eagly, A. H., and V. J. Steffen. 1986. "Gender and Aggressive Behavior: A Meta-analytic Review of the Social Psychological Literature." *Psychological Bulletin* 100 (3): 309; Archer, J. 2009. "Does Sexual Selection Explain Human Sex Differences in Aggression?" *Behavioral and Brain Sciences* 32: 249–311.

30. Dabbs, J. M., and M. G. Dabbs. 2000. *Heroes, Rogues, and Lovers: Testosterone and Behavior.* New York: McGraw-Hill; Brescoll, V. L. 2012. "Who Takes the Floor and Why: Gender, Power, and Volubility in Organizations." *Administrative Science Quarterly.*

31. Brown, S., and C. Vaughan. 2009. *Play: How It Shapes the Brain, Opens the Imagination, and Invigorates the Soul.* New York: Avery Penguin Group.

32. Vygotsky, L. S. 1978. *Mind in Society.* Cambridge, MA: Harvard University Press.

ABOUT THE EDITORS

For more than twenty years, **Dr. Norbert Kraft** has worked in aviation and aerospace research and development. His primary area of expertise is developing physiological and psychological countermeasures to combat the negative effects of long-duration spaceflight. And since 2012, Dr. Kraft has served as the Chief Medical Officer of Mars One, whose aim is to send the first people to Mars in 2027. Dr. Kraft is responsible for the selection, training, and medical–physiological–psychological well-being of all Mars One astronauts.

Dr. Kraft's work experiences span Europe, Asia, and America. Early in his career, he designed standards for using bicycle ergometry to assess cardiovascular parameters and developed a health risk score and training program for pilots and patients with cardiological problems. He improved the rehabilitation machine "Meditrain," an enhancement method and therapy for muscle and neurological problems and a countermeasure for bone demineralization and muscle de-conditioning due to microgravity.

From 1997 to 2002, Dr. Kraft worked for the Japanese Space Agency (JAXA), where he designed and supported JAXA's first isolation chamber project for medical and psychological research and was the first to incorporate isolation chamber results into astronaut selection. He was actively involved in the selection of three Japanese astronauts.

In 2000, Dr. Kraft was Principal Investigator during a 278-day spaceflight simulation experiment performed in isolation in Moscow, Russia. Furthermore, he joined the experiment as a participant for 110 days, during which time he was the Commander (selected by the Russian Space Agency) of the international mixed-gender crew.

For ten years at NASA, Dr. Kraft developed new approaches to enhance team performance for exploration missions and to enable distributed crews to respond to unanticipated problems and collaborate effectively under task stressors associated with space missions. His research at NASA includes identifying biomedical correlates of psychosocial decrements affecting team performance, and evaluating and improving physiological monitoring tools. He drove groundbreaking research efforts to study effects of prolonged isolation and confinement on the mental health of astronauts, to ultimately enable prolonged habitation of the International Space Station and space exploratory missions to Mars and beyond.

Dr. Kraft has also worked in investigating and developing countermeasures for work schedule–related alertness decrements of FAA Air Traffic Controllers and in pilots for a major South American airline. Dr. Kraft cofounded the Mental and Physical Performance Improvement Group in 2008 with an office in Lima, Peru (www.mpigp.com).

In 2013, Dr. Kraft received the NASA Group Achievement Award, one of the most prestigious awards a group can receive, which is presented to selected groups who have distinguished themselves by making outstanding contributions to the NASA mission. The award was given for "NASA Controller and Fatigue Monitoring study in evaluating the effectiveness of schedule changes for FAA Air Traffic Services." His research evaluated Traffic Controller alert and fatigue monitoring as well as cognitive decrements during short- and long-haul flights in pilots and the physiological and psychological workload of flight attendants on long-haul flights. His findings may lead to changes in flight-schedule regulations.

In 2010, Dr. Kraft received the Aerospace Medical Association Raymond F. Longacre Award for Outstanding Accomplishments in the Psychological and Psychiatric Aspects of Aerospace Medicine.

Dr. Kraft is an author of over forty papers in the field of aerospace medicine, including a seminal paper on intercultural crew issues in long-duration spaceflight, "Psychology and Culture during Long Duration Space Missions," published in the journal *Aviation, Space, and Environmental Medicine*, and was coauthor of the book *On Orbit and Beyond: Psychological Perspectives on Human Spaceflight.*

Dr. James R. Kass holds a BSc in physics and math (Montreal) and an MS in physics (Ann Arbor, Michigan). He completed his PhD in physics in Leeds, England, before joining the Max-Planck Institute to research in nuclear physics. He then joined the department of medicine at Johannes Gutenberg University, Mainz, Germany, where he did research in neurophysiology and space medicine for eight years, working on two Spacelab missions. He coordinated the European SpaceSled experimenters teams and was responsible for crew training, crew procedures, and ground operations, communicating directly with the space crew.

Following this academic work he joined the aerospace industry in Bremen and Munich in the sectors of space operations, telemedical research, artificial intelligence, and human behavior, performing pilot experiments underwater and in parabolic flight, and testing and validating procedures, experiments, and telemedical equipment running on the Mir station during the Mir97 mission. He has also worked intimately with cosmonauts of the Russian Salyut space station, as well as with astronauts of the Apollo program, Skylab, Shuttle, and ISS. In addition, he was involved in several space-simulation isolation studies in Cologne, Moscow, and Toronto.

In 2000, he joined the European Space Agency, directorate of human spaceflight and microgravity. He then worked in several other directorates and departments before setting up his own consultancy in 2010.

Over the years Dr. Kass has taught at various academic institutions: He has been Adjunct Associate Professor at Concordia University, Montreal, Department of Applied Human Sciences, where he lectured and carried out research in human behavior; he has lectured at the International Space University in Strasbourg and Barcelona (summer session); and has lectured for more than ten years at University College London, Centre for Altitude, Space, and Extreme Environment.

As a spin-off of this work in human spaceflight, Dr. Kass worked in the domain of telemedicine and eHealth. He was a founding member from 2002 of the Telemedicine Alliance, a collaborative endeavor involving the European Space Agency (ESA), the World Health Organization (WHO), the International Telecommunication Union (ITU), and the European Commission; the work covered almost all aspects of ICT in health care (eHealth), bringing Dr. Kass into close relationship with government agencies and ministries across Europe, as well as with standardization bodies, for which he has been called to advise and participate in expert working groups.

More recently, Dr. Kass helped launch ESA's Integrated Applications Promotion program, where innovative applications based on Earth observation, telecommunications, and navigation satellites are designed and tested for potential commercial operations.

Learning lessons from experience has always been of paramount importance to Dr. Kass during this long multidisciplinary professional career. Putting this into practice, he set up a Lessons Learned system at ESA, and currently advises the European Commission in this important domain.

Dr. Kass has authored numerous publications in scientific and technical journals, and made presentations and given keynote speeches at international scientific symposia, congresses, workshops, trade fares, universities, and intergovernmental agencies.

Dr. Raye Kass is a Professor of Applied Human Sciences at Concordia University (Montreal, QC). Specializing in leadership and small-group behavior, she is the author of *Theories of Small Group Development* and coauthor of three other books.

Dr. Kass has led more than a thousand workshops in leadership training, communication and problem-solving skills, staff development, team building, and conflict management with various universities, research centers, schools, hospitals, penitentiaries, government departments, and organizations in Canada, the United States, Europe, and Asia.

Dr. Kass has been highlighted frequently by both national and international press agencies for both her space sciences and group theory research. Dr. Kass was Principal Investigator on two space-simulation missions: the 1994 CAPSULS Mission held in Canada and the 240-day space simulation SFINCSS Mission held in Russia from 1999 to 2000. Her ground-based research project with the NASA Ames Research Center examined the effectiveness of various training approaches to counteract team dysfunction among multicultural and gender-mixed teams.